Nonlinear analysis and mechanics: Heriot-Watt Symposium

VOLUME 1

R J Knops (Editor)
Heriot-Watt University

Nonlinear analysis and mechanics: Heriot-Watt Symposium

VOLUME I

Pitman
LONDON · SAN FRANCISCO · MELBOURNE

PITMAN PUBLISHING LIMITED
39 Parker Street, London WC2B 5PB

FEARON-PITMAN PUBLISHERS INC.
6 Davis Drive, Belmont, California 94002, USA

Associated Companies
Copp Clark Ltd, Toronto
Pitman Publishing Co. SA (Pty) Ltd, Johannesburg
Pitman Publishing New Zealand Ltd, Wellington
Pitman Publishing Pty Ltd, Melbourne

First published 1977

AMS Subject Classifications: (main) 34–XX
(subsidiary) 46–XX, 73C10, 35B40

British Library Cataloguing in Publication Data

Nonlinear analysis and mechanics.–(Research notes in mathematics; no. 17).
1. Elasticity–Congress 2. Differential equations, partial–Congresses 3. Nonlinear mechanics–Congresses
I. Knops, R J II. Series
531'.2832 QA931

ISBN 0-273-01128-6

Reproduced and printed by photolithography
in Great Britain at Biddles of Guildford.

Preface

This volume consists mainly of written versions of invited lectures given at two short symposia held in May and September 1976 at Heriot-Watt University. The symposia formed part of a Science Research Council sponsored research programme into "Qualitative Properties of Nonlinear Elasticity", and the aim of the lectures was to describe recent methods and results in associated areas of differential equations, analysis and mechanics that might be helpful in furthering the objectives of the programme. Thus, the topics treated are not restricted to nonlinear elasticity and may be of value to those whose immediate interests do not lie in that subject.

The task of organizing the material for publication has been made easy by the thoroughness with which each author has converted the spoken to the written word. Thanks are due to them not only for this and their continual cooperation, but also for their agreement to lecture in the first place. It is also a pleasure to acknowledge the skill and patience of Mrs. Linda I. W. Gamble who accurately typed all the manuscripts, and of Pauline Houville who prepared the diagrams.

Further volumes are planned in the same series based on further lectures given as part of the S.R.C. research programme.

Edinburgh
July 1977

R. J. Knops

Contents

Preface

C. M. Dafermos

CHARACTERISTICS IN HYPERBOLIC CONSERVATION LAWS. A STUDY OF THE STRUCTURE AND THE ASYMPTOTIC BEHAVIOR OF SOLUTIONS

1 Introduction 1
2 Admissible Solutions 6
3 Characteristics 7
4 Structure of Solutions 17
5 Smoothness of Solutions 26
6 A Useful Identity 34
7 Decay of Solutions 39
8 Initial Conditions Asymptotic to Riemann Data. The Shock Case 42
9 Initial Conditions Asymptotic to Riemann Data. The Wave Case 46
10 More Precise Results on Asymptotic Behaviour of Solutions 51
References 56

J. K. Hale

GENERIC BIFURCATION WITH APPLICATIONS

1 Introduction 59
2 Method of Liapunov-Schmidt and Alternative Problems 66
3 $\dim \mathfrak{N}(B) = 1 = \operatorname{codim} \mathcal{R}(B)$, $\dim \Lambda = 1$ 70
4 Normal Eigenvalues of Odd Dimensions 76
5 $\dim \mathfrak{N}(B) = 1 = \operatorname{codim} \mathcal{R}(B)$, $\dim \Lambda > 1$ 78
6 Some Applications 84
7 Normal Forms, $\dim \mathfrak{N}(B) = 1 = \operatorname{codim} \mathcal{R}(B)$ 92
8 Generalised Hopf Bifurcation 96
9 The Cusp Revisited 98
10 Bifurcation in Two Dimensions. Cubic Nonlinearities and Two Parameters 103

11 Bifurcation in Two Dimensions. Quadratic Nonlinearities and Two Parameters 108
12 Applications to a Special Family in Banach Space 110
13 Applications to Rectangular Plates 112
14 A Three Parameter Family in the Buckling of Plates 116
15 Buckling of Cylindrical Shells with Small Curvature 125
16 Some Remarks on the Generic Hypotheses 132
17 Bifurcation from Families of Solutions - Applications to Nonlinear Oscillations 135
18 Harmonics in the Duffing Equation 143
References 151

J. L. Ericksen

ON THE FORMULATION OF ST.-VENANT'S PROBLEM

1 Introduction 158
2 Basic Equations 160
3 Kinematics 164
4 Equilibrium Equations 169
5 Torsion 175
6 General Considerations 179
References 186

J. M. Ball

CONSTITUTIVE INEQUALITIES AND EXISTENCE THEOREMS IN NONLINEAR ELASTOSTATICS

1 Introduction 187
2 Constitutive Inequalities 203
3 Null Lagrangians and Sequentially Weakly Continuous Functions 214
4 Existence of Minimizers 221
5 Remarks on Regularity and Stored-Energy Functions of Slow Growth 225
6 Applications to Specific Elastic Materials 229
7 Existence of Semi-Inverse Solutions 233
References 238

C M DAFERMOS

Characteristics in hyperbolic conservation laws. A study of the structure and the asymptotic behaviour of solutions

In these lecture notes we discuss certain aspects of the theory of the conservation law

$$u_t + f(u)_x = 0 \tag{1.1}$$

by a new approach. The function $f(\cdot)$ is C^2 smooth and is strictly (but not necessarily uniformly) convex, i.e., $f''(\cdot)$ is nonnegative and does not vanish identically on any interval.

The study of the qualitative theory of (1.1) was initiated by Hopf [1] for the case $f(u) = \frac{1}{2}u^2$. Subsequently, Lax [2] discovered an explicit representation of solutions of the initial-value problem for (1.1) with arbitrary convex $f(\cdot)$. Several other methods have been developed for the study of (1.1), based on a priori estimates (e.g., [3,4,5,6,7]). The advantage of these methods is that they are also applicable to the nonhomogeneous equation

$$u_t + f(u,x,t)_x + g(u,x,t) = 0, \tag{1.2}$$

with or without convexity of $f(\cdot,x,t)$, as well as to the case of several space variables. On the other hand, for (1.1), the explicit solution of Lax still is the source of the most precise available information.

In addition to existence and uniqueness, the aspects of the theory of solutions of (1.1) that have been investigated so far pertain to structure and asymptotic behavior. It is well-known that, in general, the initial-value problem for (1.1) does not have global smooth solutions even if the initial data are very smooth. Classical prototypes of weak solutions are constructed in the class of piecewise smooth functions with jump discontinuities across smooth curves. In continuum physics, which has motivated the study of conservation laws, solutions in this function class admit a natural interpretation, the lines of discontinuity being interpreted as trajectories of propagating shock waves. Unfortunately, the class of piecewise smooth functions is too narrow to encompass all solutions of (1.1), even those with C^∞ smooth initial data.

Research on hyperbolic conservation laws has revealed that the natural framework in which solutions should be studied is provided by the class of functions of bounded variation in the sense of Tonelli and Cesari(*). Volpert [5] has called attention to the analogy between the geometric structure of piecewise smooth functions and that of functions of bounded variation, with points of smoothness and jump discontinuity in the former class corresponding to points of approximate continuity and approximate jump discontinuity in the latter class. This structure is inherited by solutions of (1.1). The natural question, of course, is whether membership in the class of functions of bounded variation provides maximal information on the structure of solutions.

Using an explicit representation of solutions (similar to Lax's, cited above), Oleinik [8,9,3] studies the structure of solutions of (1.1) (as well as of (1.2) in the special case $g \equiv 0$) and shows that solutions are continuous except on the union of an at most countable set of Lipschitz continuous curves (shocks). Analogous results were established by DiPerna [10] for solutions of genuinely nonlinear systems of conservation laws constructed by Glimm's scheme [11,12].

(*)The relevance of this class was first recognized by Conway and Smoller [4]. The importance of estimates involving variation in the study of (1.1) had been demonstrated earlier by Oleinik (e.g. [3]).

The above results do not preclude the possibility of solutions with quite complicated structure, e.g., with shock sets that are everywhere dense on the domain of definition. Nevertheless, Schaeffer [13], combining Thom's theory of catastrophes with Lax's representation, shows that generically solutions with smooth initial data are piecewise smooth (see also [14]).

In contrast to the approaches cited above, which are associated with specific construction schemes for solutions, our point of view here is to study solutions of (1.1) a priori. Namely, we consider a solution in an appropriate function class introduced in Section 2 (somewhat broader than the class of functions of bounded variation), with no reference to how this solution was constructed, and we proceed to study its structure.

The main tool in implementing our program is the concept of a generalized characteristic. Recall that a characteristic of (1.1) associated with a classical solution $u(x,t)$ is defined as a trajectory of the ordinary differential equation

$$\frac{dx}{dt} = f'(u(x,t)). \tag{1.3}$$

Classical characteristics are straight lines on which u is constant. For weak solutions the same definition of a characteristic is adopted here where, however, (1.3) is now interpreted as a contingent equation in the sense of Filippov [15]. In Section 3 we show that a generalized characteristic is either a shock or a classical characteristic, i.e., a straight line on which u is constant. From each point (x,t) of the upper half-plane there emanates a unique forward characteristic. On the other hand, the set of backward characteristics through (x,t) either consists of a single classical characteristic or it contains an infinite number of characteristics spanning the sector confined between two classical characteristics.

Using the above information we proceed in Section 4 to the study of the structure of solutions. We prove that each shock is a Lipschitz continuous curve which is differentiable except at points of interaction with other shocks or centered compression waves. The set of shocks is at most countable and the solution is continuous on the complement of the shock set and Lipschitz continuous in the interior of the above set.

In Section 5 we study the smoothness of solutions with smooth initial data. We show that the complement of the shock set is open and the solution is smooth on this set. We are also able to characterize, by elementary methods, the set of smooth initial data that may generate solutions which are not piecewise smooth and to show that this set is of the first category. Thus, solutions of (1.1) with smooth initial data are generically piecewise smooth.

The results obtained here, in Sections 3 - 5, for (1.1) are also established, by similar techniques, in [16] for the more general equation (1.2), with $f(\cdot,x,t)$ convex. The reason we discuss here the particular case (1.1) in detail is that its special structure and, in particular, the property that classical characteristics are straight lines makes the proofs considerably simpler and geometrically transparent.

In Section 6 we use the theory of characteristics to establish a new estimate for pairs of solutions of (1.1). This yields most of the known a priori estimates for solutions of (1.1) and, in addition, has interesting implications on the asymptotic behavior of solutions that are explored in Sections 7 - 10.

The asymptotic behavior of solutions of (1.1) is studied systematically by Lax [2] who generalizes earlier work by Hopf [1] and other authors. (For various extensions see [17,12,18]). In Section 7 we prove a proposition that unifies and generalizes many of the previously known results by relating the rate of decay of solutions to the rate of growth of the primitive of the initial data and the "strictness" of convexity of $f(\cdot)$ at the origin.

In Sections 8 and 9 we discuss the asymptotic behavior of solutions with initial conditions asymptotic to Riemann data. This problem was also studied by Ilin and Oleinik [19,20] by the viscosity method but our results here are more precise.

The final Section 10 deals with solutions with initial data that are periodic or have compact support. In these cases very accurate description of the asymptotic behavior can be given.

Acknowledgment The research reported here was done for the most part while I was visiting Heriot-Watt University, Edinburgh, under a research program sponsored by the Science Research Council of Great Britain. I am grateful to John Ball, Jack Carr and Robin Knops for their warm hospitality and for giving me the opportunity to work in a very stimulating environment. I benefited very much from discussions with them. I am equally indebted to Ronald DiPerna for many valuable ideas and discussions. The research was also supported by the U.S. Army under contract AROD AAG 29-76-6-0052, by the Office of Naval Research under contract NONR NOOO 14-75-C-0278 A04 and by the National Science Foundation under grant MPS 71-029.

2 ADMISSIBLE SOLUTIONS

We first introduce the class of functions in which admissible solutions will be studied.

Definition 2.1 A measurable, locally bounded function $u(x,t)$ on $(-\infty,\infty)\times[0,\infty)$ is of class $\mathscr{K}$ if for almost all $t\in[0,\infty)$ one-sided limits $u(x-,t)$, $u(x+,t)$ exist for all $x\in(-\infty,\infty)$.

Remark 2.1 Functions of locally bounded variation in the sense of Tonelli and Cesari [5] are of class $\mathscr{K}$. In particular, if $u(x,t)$ is of locally bounded variation, then $u(\cdot,t)$ is of locally bounded variation on $(-\infty,\infty)$ for almost all (fixed) $t\in[0,\infty)$.

Definition 2.2 A function $u(x,t)$ of class $\mathscr{K}$ which satisfies (1.1) in the sense of distributions is an admissible solution if for almost all $t\in[0,\infty)$,

$$u(x-,t)\geqslant u(x+,t) \tag{2.1}$$

for all $x\in(-\infty,\infty)$.

For a discussion on the motivation of the admissibility condition (2.1) and other analogous conditions we refer, e.g., to [21].

3 CHARACTERISTICS

Throughout this section u(x,t) will denote an admissible solution of (1.1).

Definition 3.1 A Lipschitz continuous curve $\xi(\cdot): [a,b] \longrightarrow (-\infty,\infty)$ is called a characteristic if for almost all $t \in [a,b]$

$$\dot{\xi}(t) \in [f'(u(\xi(t)+,t)) , f'(u(\xi(t)-,t))]. \tag{3.1}$$

By the theory of contingent equations of type (3.1) [15], through any fixed point $(\bar{x},\bar{t}) \in (-\infty,\infty)\times(0,\infty)$ there is at least one forward characteristic, defined on a maximal interval $[\bar{t},s)$, and at least one backward characteristic, defined on a maximal interval $(\sigma,\bar{t}]$. The set of forward (or backward) characteristics through $(\bar{x},\bar{t})$ spans the funnel confined between a minimal and a maximal forward (or backward) characteristic through $(\bar{x},\bar{t})$. Of course the minimal and maximal forward (or backward) characteristics are not necessarily distinct. The minimal and maximal backward characteristics will play a central role in our discussion and will be denoted throughout by $\zeta_-(t;\bar{x},\bar{t})$ and $\zeta_+(t;\bar{x},\bar{t})$, respectively.

It may seem that (3.1) allows considerable freedom in the speed of propagation of characteristics. The following proposition shows, however, that characteristics must propagate either at classical characteristic speed or at shock speed.

<u>Theorem 3.1</u> Let $\xi(\cdot): [a,b] \longrightarrow (-\infty,\infty)$ be a characteristic. Then for almost all $t \in [a,b]$

$$\dot{\xi}(t) = \begin{cases} f'(u(\xi(t)\pm,t)) & \text{if } u(\xi(t)-,t) = u(\xi(t)+,t) \\ \dfrac{f(u(\xi(t)+,t))-f(u(\xi(t)-,t))}{u(\xi(t)+,t)-u(\xi(t)-,t)} & \text{if } u(\xi(t)-,t) > u(\xi(t)+,t). \end{cases} \tag{3.2}$$

Theorem 3.1 is an immediate corollary of (3.1) and Lemma 3.1, below.

<u>Lemma 3.1</u> Let $\xi(\cdot): [a,b] \longrightarrow (-\infty,\infty)$ be a Lipschitz continuous curve, $0 \leqslant a < b < \infty$. Then, for almost all $t \in [a,b]$,

$$f(u(\xi(t)+,t))-f(u(\xi(t)-,t))-\dot{\xi}(t)[u(\xi(t)+,t)-u(\xi(t)-,t)] = 0. \tag{3.3}$$

<u>Lemma 3.2</u> Let $\xi(\cdot): [a,b] \longrightarrow (-\infty,\infty)$ and $\zeta(\cdot): [a,b] \longrightarrow (-\infty,\infty)$ be Lipschitz continuous curves, $0 \leqslant a < b < \infty$. Then, for almost all σ,τ with $a \leqslant \sigma < \tau \leqslant b$,

$$\int_{\xi(\tau)}^{\zeta(\tau)} u(x,\tau)dx - \int_{\xi(\sigma)}^{\zeta(\sigma)} u(x,\sigma)dx$$
$$= \int_{\sigma}^{\tau}[f(u(\xi(t)-,t))-\dot{\xi}(t)u(\xi(t)-,t)]dt- \int_{\sigma}^{\tau}[f(u(\zeta(t)+,t))-\dot{\zeta}(t)u(\zeta(t)+,t)]dt. \tag{3.4}$$

<u>Proof of Lemmas 3.1 and 3.2.</u> Assume first that $\xi(t) \leqslant \zeta(t)$, $t \in [a,b]$. Since $u(x,t)$ is a solution of (1.1) in the sense of distributions,

$$\int_{0}^{\infty}\int_{-\infty}^{\infty} [u\phi_t + f(u)\phi_x]dxdt = 0 \tag{3.5}$$

for any Lipschitz continuous test function $\phi(x,t)$ defined on $(-\infty,\infty)\times(-\infty,\infty)$

and having compact support contained in $(-\infty,\infty)\times[0,\infty)$.

We pick σ and τ in the set(*) of points of approximate continuity of the family of functions

$$\int_c^d u(x,t)dx, \quad -\infty < c < d < \infty.$$

For ε positive small we define

$$\phi_\varepsilon(x,t) = \psi_\varepsilon(x,t)\,\chi_\varepsilon(t), \tag{3.6}$$

where

$$\chi_\varepsilon(t) = \begin{cases} 0 & t < \sigma \\ \dfrac{t-\sigma}{\varepsilon} & \sigma \leqslant t < \sigma + \varepsilon \\ 1 & \sigma + \varepsilon \leqslant t < \tau \\ 1 + \dfrac{\tau - t}{\varepsilon} & \tau \leqslant t < \tau + \varepsilon \\ 0 & \tau + \varepsilon \leqslant t \end{cases} \tag{3.7}$$

$$\psi_\varepsilon(x,t) = \begin{cases} 0 & x < \xi(t) - \varepsilon \\ 1 + \dfrac{x-\xi(t)}{\varepsilon} & \xi(t)-\varepsilon \leqslant x < \xi(t) \\ 1 & \xi(t) \leqslant x \leqslant \zeta(t) \\ 1 + \dfrac{\zeta(t)-x}{\varepsilon} & \zeta(t) < x \leqslant \zeta(t) + \varepsilon \\ 0 & \zeta(t) + \varepsilon < x. \end{cases} \tag{3.8}$$

Applying (3.5) for the test function $\phi_\varepsilon(x,t)$, letting $\varepsilon \longrightarrow 0+$, and recalling that $u(x,t)$ is of class $\mathscr{K}$ (Definition 2.1) we arrive at (3.4) for the case $\xi(t) \leqslant \zeta(t)$, $t \in [a,b]$.

In particular, applying (3.4) with $\zeta(t) \equiv \xi(t)$, we obtain (3.3) thus proving Lemma 3.1.

(*) This set has total measure in $[0,\infty)$.

In the general case where $\xi(t) - \zeta(t)$ may become positive at some (or at all) points of $[a,b]$, we first write (3.4) for the functions $\bar{\xi}(t) = \min\{\xi(t),\zeta(t)\}$, $\bar{\zeta}(t) = \max\{\xi(t),\zeta(t)\}$ and then use Lemma 3.1 to derive (3.4) for $\xi(\cdot)$ and $\zeta(\cdot)$. The proof is complete.

From Lemma 3.2 we deduce immediately the following

Corollary 3.1 After, possibly, a modification of $u(x,t)$ on a set of measure zero, $u(\cdot,t)$ is locally bounded and measurable for all $t \in [0,\infty)$ and $t \longmapsto u(\cdot,t)$, as a map from $[0,\infty)$ to $L^1_{\ell oc}(-\infty,\infty)$, is weakly continuous, i.e., for any $-\infty < c < d < \infty$, the function

$$\int_c^d u(x,t)dx$$

is continuous (and as a matter of fact Lipschitz continuous) on $[0,\infty)$. Furthermore, (3.4) is satisfied for all σ,τ in $[a,b]$.

Theorem 3.1 shows that generalized characteristics in the sense of Definition 3.1 are generalized characteristics in the sense of Glimm and Lax [12]. Within the framework of piecewise smooth solutions, generalized characteristics are classical characteristics or shock waves or combinations thereof. The (entropy) shock admissibility condition (2.1) allows (and actually induces) classical characteristics to run into shocks but prohibits classical characteristics from branching out of shocks. It is our program here to establish similar results for admissible solutions of class $\mathscr{K}$. As a first step in this direction and in order to distinguish classical characteristics from shocks within the framework of weak solutions we introduce the following

Definition 3.2 A characteristic $\xi(\cdot): [a,b] \longrightarrow (-\infty,\infty)$ is called genuine if $u(\xi(t)-,t) = u(\xi(t)+,t)$ for almost all $t \in [a,b]$.

The following proposition establishes the existence of genuine characteristics.

Theorem 3.2 For any $(\bar{x},\bar{t}) \in (-\infty,\infty)\times(0,\infty)$, the minimal and maximal backward characteristics $\zeta_-(t;\bar{x},\bar{t})$ and $\zeta_+(t;\bar{x},\bar{t})$ through $(\bar{x},\bar{t})$ are genuine.

Proof. We abbreviate $\zeta_-(t;\bar{x},\bar{t})$ by $\zeta(t)$ and we prove that it is genuine on its domain $(s,\bar{t}]$. If $\zeta(\cdot)$ is not genuine, we can find a measurable set $J, \bar{J} \subset (s,\bar{t}]$, with $\mu(J) > 0$(*), and a number $\beta > 0$ such that $u(\zeta(t)-,t)-u(\zeta(t)+,t) > \beta$ for $t \in J$. Since $f(\cdot)$ is strictly convex, there is $\varepsilon > 0$ such that for $t \in J$

$$f'(u(\zeta(t)-,t)) - \frac{f(u(\zeta(t)+,t))-f(u(\zeta(t)-,t))}{u(\zeta(t)+,t)-u(\zeta(t)-,t)} > 2\varepsilon. \tag{3.9}$$

For each $t \in J$ there exists $\delta(t) > 0$ with the property

$$f'(u(x+,t)) \geqslant f'(u(\zeta(t)-,t))-\varepsilon \ , \ \zeta(t)-\delta(t) < x < \zeta(t). \tag{3.10}$$

Finally, there is a subset I of J with $\mu^*(I) > 0$ and $\bar{\delta} > 0$ such that $\delta(t) \geqslant \bar{\delta}$ for $t \in I$.

(*)Here and throughout μ and μ^* denote, respectively, the one-dimensional Lebesgue measure and outer measure.

Let τ be a density point of I, with respect to μ^*. We can thus find $\bar{r}$, $0 < \bar{r} < \bar{t} - \tau$, so that

$$\frac{\mu^*(I \cap [\tau,\tau+r])}{r} > \frac{2|\alpha| + \varepsilon}{2|\alpha| + 2\varepsilon}, \quad 0 < r \leqslant \bar{r}, \tag{3.11}$$

where

$$\alpha \overset{\text{def}}{=} \inf \{f'(u(x+,t))-f'(u(\zeta(t)-,t)) \,\big|\, 0 \leqslant t \leqslant \bar{t}, \zeta(t)-\bar{\delta} \leqslant x < \zeta(t)\}. \tag{3.12}$$

We now fix a point $y \in (\zeta(\tau) - \bar{\delta}, \zeta(\tau))$ with

$$y > \zeta(\tau) - \frac{\varepsilon\bar{r}}{2}, \tag{3.13}$$

and we consider a forward characteristic $\xi(\cdot)$ through (y,τ). We first observe that since, by hypothesis, $\zeta(\cdot)$ is the minimal backward characteristic through $(\bar{x},\bar{t})$,

$$\xi(t) < \zeta(t), \quad t > \tau, \tag{3.14}$$

and we proceed to show that this leads to a contradiction.

We begin by showing that $\xi(\cdot)$ does not escape on $[\tau,\tau+\bar{r}]$ and

$$\xi(t) > \zeta(t) - \bar{\delta}, \quad t \in [\tau,\tau + \bar{r}]. \tag{3.15}$$

Indeed, suppose that for some $r \in (0,\bar{r}]$, $\xi(t) > \zeta(t) - \bar{\delta}$, $t \in [\tau,\tau + r)$, but $\xi(\tau + r) = \zeta(\tau + r) - \bar{\delta}$. Then, using (3.1), (3.2), (3.9), (3.10), (3.11) and (3.12) we obtain

$$0 = \xi(\tau+r)-\zeta(\tau+r)+\bar{\delta} = y + \int_\tau^{\tau+r} \dot{\xi}(t)dt - \zeta(\tau) - \int_\tau^{\tau+r} \dot{\zeta}(t)dt + \bar{\delta}$$

$$> \varepsilon\, \mu^*(I \cap [\tau,\tau+r]) + \alpha[r-\mu^*(I \cap [\tau,\tau+r])] > 0,$$

namely a contradiction that verifies (3.15).

On account of (3.14) and (3.15), using again (3.1), (3.2), (3.9), (3.10), (3.11), (3.12) and (3.13),

$$0 > \xi(\tau+\bar{r}) - \zeta(\tau+\bar{r}) = y + \int_{\tau}^{\tau+\bar{r}} \dot{\xi}(t)dt - \zeta(\tau) - \int_{\tau}^{\tau+\bar{r}} \dot{\zeta}(t)dt$$

$$> \varepsilon\mu^*(I \cap [\tau,\tau+\bar{r}]) + \alpha[\bar{r} - \mu^*(I \cap [\tau,\tau+\bar{r}])] - \frac{\varepsilon\bar{r}}{2} > 0,$$

and this contradiction completes the proof of the assertion that $\zeta_-(t;\bar{x},\bar{t})$ is genuine. The proof that $\zeta_+(t;\bar{x},\bar{t})$ is also genuine is similar and will thus be omitted.

The proof of Theorem 3.2 does not depend essentially on the fact that we are dealing here with a single conservation law but only on the observation that the convexity of $f(\cdot)$ together with the admissibility condition (2.1) imply that the classical characteristic speed on the left (or right) side of a shock is greater (or less) than the shock speed [2]. An analogous result will thus hold for general genuinely nonlinear systems of hyperbolic conservation laws. For conservation laws that are not genuinely nonlinear, the minimal and maximal backward characteristics shall be composed of genuine characteristics and/or contact discontinuities.

In contrast, the next proposition which asserts that genuine characteristics are classical characteristics, relies heavily on the special structure of equation (1.1) as well as on the convexity hypothesis on $f(\cdot)$.

Theorem 3.3 Let $\xi(\cdot)$ be a genuine characteristic on $[a,b]$. Then there is a constant $\bar{u}$ such that $\xi(\cdot)$ is a straight line with slope $f'(\bar{u})$. Furthermore,

$$\lim_{\varepsilon\to 0+} \frac{1}{\varepsilon} \int_{\xi(t)-\varepsilon}^{\xi(t)} u(x,t)dx = \bar{u} = \lim_{\varepsilon\to 0+} \frac{1}{\varepsilon} \int_{\xi(t)}^{\xi(t)+\varepsilon} u(x,t)dx, \quad t \in (a,b), \quad (3.16)$$

$$\limsup_{\varepsilon\to 0+} \frac{1}{\varepsilon} \int_{\xi(a)-\varepsilon}^{\xi(a)} u(x,a)dx \leqslant \bar{u} \leqslant \liminf_{\varepsilon\to 0+} \frac{1}{\varepsilon} \int_{\xi(a)}^{\xi(a)+\varepsilon} u(x,a)dx, \tag{3.17}$$

$$\liminf_{\varepsilon\to 0+} \frac{1}{\varepsilon} \int_{\xi(b)-\varepsilon}^{\xi(b)} u(x,b)dx \geqslant \bar{u} \geqslant \limsup_{\varepsilon\to 0+} \frac{1}{\varepsilon} \int_{\xi(b)}^{\xi(b)+\varepsilon} u(x,b)dx. \tag{3.18}$$

In particular,

$$u(\xi(t)-,t) = \bar{u} = u(\xi(t)+,t) \ , \ \text{a.e. on } (a,b). \tag{3.19}$$

Proof. We fix σ,τ $a \leqslant \sigma < \tau \leqslant b$, and $\varepsilon > 0$. We apply (3.4) with $\zeta(t) = \xi(t)-\varepsilon$ and use the convexity of $f(\cdot)$ to deduce

$$\int_{\xi(\tau)-\varepsilon}^{\xi(\tau)} u(x,\tau)dx - \int_{\xi(\sigma)-\varepsilon}^{\xi(\sigma)} u(x,\sigma)dx = \int_{\sigma}^{\tau} \{f(u(\xi(t)-\varepsilon+,t))-f(u(\xi(t)-,t)) -$$

$$f'(u(\xi(t)-,t))[u(\xi(t)-\varepsilon+,t)-u(\xi(t)-,t)]\}dt \geqslant 0.$$

Therefore,

$$\lim_{\varepsilon\to 0+} \begin{matrix}\inf\\ \sup\end{matrix} \frac{1}{\varepsilon} \int_{\xi(\tau)-\varepsilon}^{\xi(\tau)} u(x,\tau)dx \geqslant \lim_{\varepsilon\to 0+} \begin{matrix}\inf\\ \sup\end{matrix} \frac{1}{\varepsilon} \int_{\xi(\sigma)-\varepsilon}^{\xi(\sigma)} u(x,\sigma)dx. \tag{3.20}$$

Similarly, applying (3.4) with $\zeta(t) = \xi(t)+\varepsilon$ and following the same procedure, we obtain

$$\lim_{\varepsilon\to 0+} \begin{matrix}\inf\\ \sup\end{matrix} \frac{1}{\varepsilon} \int_{\xi(\tau)}^{\xi(\tau)+\varepsilon} u(x,\tau)dx \leqslant \lim_{\varepsilon\to 0+} \begin{matrix}\inf\\ \sup\end{matrix} \frac{1}{\varepsilon} \int_{\xi(\sigma)}^{\xi(\sigma)+\varepsilon} u(x,\sigma)dx. \tag{3.21}$$

We now fix $\bar{t} \in (a,b)$ with $u(\xi(\bar{t})-,\bar{t}) = u(\xi(\bar{t})+,\bar{t}) \overset{\text{def}}{=} \bar{u}$. Applying (3.20), (3.21) with $\sigma = a$, $\tau = \bar{t}$ we obtain (3.17). Similarly, (3.20), (3.21) with $\sigma = \bar{t}$, $\tau = b$ yield (3.18). For any $t \in (a,b)$ with $u(\xi(t)-,t) = u(\xi(t)+,t)$ we apply (3.20), (3.21) with $\sigma = \bar{t}$, $\tau = t$, if $t > \bar{t}$, or $\sigma = t$, $\tau = \bar{t}$, if $t < \bar{t}$, thus arriving at (3.19). In particular, by (3.1), $\dot{\xi}(t) = f'(\bar{u})$ for almost all $t \in (a,b)$ so that $\xi(\cdot)$ is a straight line with slope $f'(\bar{u})$. Finally, in order to prove (3.16), we apply (3.20), (3.21) first with $\sigma = t$ and $\tau > t$ such that $u(\xi(\tau)-,\tau) = u(\xi(\tau)+,\tau) = \bar{u}$ and then with $\tau = t$ and $\sigma < t$ such that $u(\xi(\sigma)-,\sigma) = u(\xi(\sigma)+,\sigma) = \bar{u}$. The proof is complete.

From the above result we deduce immediately the following

Corollary 3.2 Two genuine characteristics may intersect only at their end points.

Combining Theorems 3.2 and 3.3 we conclude that the minimal and maximal backward characteristics $\zeta_-(t;\bar{x},\bar{t})$ and $\zeta_+(t;\bar{x},\bar{t})$ through any point $(\bar{x},\bar{t}) \in (-\infty,\infty) \times (0,\infty)$ are straight lines with finite slopes and, therefore, cannot escape at any point of $[0,\bar{t}]$. Thus

Corollary 3.3 For any $(\bar{x},\bar{t}) \in (-\infty,\infty) \times (0,\infty)$, the backward characteristics through $(\bar{x},\bar{t})$ are defined on $[0,\bar{t}]$.

An analogous result holds for forward characteristics, namely:

Corollary 3.4 For every $(\bar{x},\bar{t}) \in (-\infty,\infty) \times [0,\infty)$, any forward characteristic through $(\bar{x},\bar{t})$ cannot escape at a finite time and is thus defined on $[\bar{t},\infty)$.

Proof. Let $\eta(\cdot)$ be a forward characteristic through $(\bar{x},\bar{t})$ and suppose that it escapes at $s \in (\bar{t},\infty)$, say $\eta(t_n) \longrightarrow \infty$ for some sequence $\{t_n\}$ converging to $s-$. The slope of $\zeta_-(t;\eta(t_n),t_n)$ will then tend to infinity as $n \longrightarrow \infty$. Furthermore, since $\zeta_-(t;\eta(t_n),t_n) \leqslant \eta(t)$, $t \in [\bar{t},t_n]$, $\zeta_-(t;\eta(t_n),t_n)$ must enter the rectangle $\{\bar{x}-1 < x < \bar{x}+1,\ 0 < t < s\}$ which is impossible since $u(x,t)$ is bounded on this rectangle. The proof is complete.

An important consequence of Corollary 3.2 is the following

Theorem 3.4 There is a unique forward characteristic through any point $(\bar{x},\bar{t}) \in (-\infty,\infty) \times (0,\infty)$.

Proof. Suppose there were two forward characteristics $\eta(\cdot)$ and $\xi(\cdot)$ through $(\bar{x},\bar{t})$ with $\eta(\tau) < \xi(\tau)$ for some $\tau > \bar{t}$. Consider the maximal backward characteristic $\zeta_+(t;\eta(\tau),\tau)$ through $(\eta(\tau),\tau)$ and the minimal backward characteristic $\zeta_-(t;\xi(\tau),\tau)$ through $(\xi(\tau),\tau)$. For $t \in [\bar{t},\tau]$, $\eta(t) \leq \zeta_+(t;\eta(\tau),\tau)$ and $\xi(t) \geq \zeta_-(t;\xi(\tau),\tau)$. Moreover, $\eta(\bar{t}) = \xi(\bar{t}) = \bar{x}$. Therefore, $\zeta_+(t;\eta(\tau),\tau)$ and $\zeta_-(t;\xi(\tau),\tau)$ must intersect for some $t \in [\bar{t},\tau)$ and this is a contradiction to Corollary 3.2. The proof is complete.

It should be noted, however, that more than one forward characteristics may emanate from points of the axis $t = 0$.

In the next section we will employ the properties of characteristics obtained here to get information on the structure of solutions. In turn this will yield additional information on the structure of characteristics.

4 STRUCTURE OF SOLUTIONS

In this section we consider an admissible solution $u(x,t)$ of (1.1) and study the geometric structure of the set of points of continuity and discontinuity.

Theorem 4.1 After possibly a modification on a set of measure zero, not destroying the continuity of $t \longmapsto u(\cdot,t)$ obtained in Corollary 3.1, $u(x,t)$ acquires the following properties: For each fixed $(\bar{x},\bar{t}) \in (-\infty,\infty) \times (0,\infty)$ the one-sided limits $u(\bar{x}\pm,\bar{t})$ exist and

$$u(\bar{x}-,\bar{t}) \geqslant u(\bar{x}+,\bar{t}). \tag{4.1}$$

Furthermore, the minimal and maximal backward characteristics $\zeta_-(t;\bar{x},\bar{t})$ and $\zeta_+(t;\bar{x},\bar{t})$ through $(\bar{x},\bar{t})$ are straight lines with slope $f'(u(\bar{x}-,\bar{t}))$ and $f'(u(\bar{x}+,\bar{t}))$, respectively, and

$$\begin{aligned} u(\zeta_-(t;\bar{x},\bar{t})\pm,t) &= u(\bar{x}-,\bar{t}) \ , \quad t \in (0,\bar{t}), \\ u(\zeta_+(t;\bar{x},\bar{t})\pm,t) &= u(\bar{x}+,\bar{t}) \ , \quad t \in (0,\bar{t}). \end{aligned} \tag{4.2}$$

In particular, $\zeta_-(t;\bar{x},\bar{t})$ and $\zeta_+(t;\bar{x},\bar{t})$ coincide if and only if $u(\bar{x}-,\bar{t}) = u(\bar{x}+,\bar{t})$.

Proof. We fix $(\bar{x},\bar{t}) \in (-\infty,\infty) \times (0,\infty)$. Assume first that $\bar{x}$ is a point of approximate continuity of the function $u(\cdot,\bar{t})$ so that

$$\lim_{\varepsilon\to 0+} \frac{1}{\varepsilon} \int_{\bar{x}-\varepsilon}^{\bar{x}} u(x,\bar{t})dx = u(\bar{x},\bar{t}) = \lim_{\varepsilon\to 0+} \frac{1}{\varepsilon} \int_{\bar{x}}^{\bar{x}+\varepsilon} u(x,\bar{t})dx. \tag{4.3}$$

Then, by Theorem 3.3 (see in particular (3.18)), both $\zeta_-(t;\bar{x},\bar{t})$ and $\zeta_+(t;\bar{x},\bar{t})$ are straight lines with slope $f'(u(\bar{x},\bar{t}))$ and, therefore, coincide.

Suppose now $(\bar{x},\bar{t})$ is arbitrary. Consider any strictly increasing sequence $\{x_n\}$, $x_n \longrightarrow \bar{x}-$, such that x_n is a point of approximate continuity of $u(\cdot,\bar{t})$, $n = 1,2, \ldots$. By Corollary 3.2,

$$\zeta_\pm(t;x_1,\bar{t}) < \zeta_\pm(t;x_2,\bar{t}) < \ldots < \zeta_-(t;\bar{x},\bar{t}), \quad t \in (0,\bar{t}]. \tag{4.4}$$

The theory of contingent equations of type (3.1) [15] asserts that $\{\zeta_\pm(t;x_n,\bar{t})\}$ must converge, uniformly on $[0,\bar{t}]$, to a backward characteristic through $(\bar{x},\bar{t})$. Since $\zeta_-(t;\bar{x},\bar{t})$ is the minimal backward characteristic through $(\bar{x},\bar{t})$ and by virtue of (4.4) we conclude that $\zeta_\pm(t;x_n,\bar{t}) \longrightarrow \zeta_-(t;\bar{x},\bar{t})$. Consequently, the sequence $\{f'(u(x_n,\bar{t}))\}$ of slopes must be convergent with limit the slope of $\zeta_-(t;\bar{x},\bar{t})$. If we thus modify appropriately $u(\cdot,\bar{t})$ on the set (of zero one-dimensional Lebesgue measure) of x that are not points of approximate continuity of $u(\cdot,\bar{t})$, we can guarantee that $u(\bar{x}-,\bar{t})$ exists and the slope of $\zeta_-(t;\bar{x},\bar{t})$ equals $f'(u(\bar{x}-,\bar{t}))$. Similarly one shows that $u(\bar{x}+,\bar{t})$ exists and $f'(u(\bar{x}+,\bar{t}))$ is the slope of $\zeta_+(t;\bar{x},\bar{t})$. In particular, $f'(u(\bar{x}-,\bar{t})) \geqslant f'(u(\bar{x}+,\bar{t}))$ and this implies (4.1). Finally, (4.2) follows by virtue of Theorem 3.3. The proof is complete.

<u>Corollary 4.1</u> For any fixed $t \in (0,\infty)$, the set of points of discontinuity of $u(\cdot,t)$ is at most countable.

We note that on account of Theorem 4.1 any genuine characteristic can be extended backwards, as a genuine characteristic, up to $t = 0$, i.e., all genuine characteristics emanate from the axis $t = 0$.

We now turn our attention to characteristics that are not genuine.

Theorem 4.2 Let $\eta(\cdot)$ be a characteristic on $[\bar{t},\infty)$, $\bar{t} > 0$. If $u(\eta(\bar{t})-,\bar{t}) > u(\eta(\bar{t})+,\bar{t})$, then $u(\eta(t)-,t) > u(\eta(t)+,t)$ for all $t \geq \bar{t}$ and $u(\eta(t)-,t) - u(\eta(t)+,t)$ is bounded away from zero, uniformly on bounded subsets of $[\bar{t},\infty)$.

Proof. Since $u(\eta(\bar{t})-,\bar{t}) > u(\eta(\bar{t})+,\bar{t})$, $\zeta_-(0;\eta(\bar{t}),\bar{t}) < \zeta_+(0;\eta(\bar{t}),\bar{t})$. On account of Corollary 3.2, for any $t \geq \bar{t}$, $\zeta_-(0;\eta(t),t) \leq \zeta_-(0;\eta(\bar{t}),\bar{t}) < \zeta_+(0;\eta(\bar{t}),\bar{t}) \leq \zeta_+(0;\eta(t),t)$. Consequently, the difference of the slopes of $\zeta_-(\tau;\eta(t),t)$ and $\zeta_+(\tau;\eta(t),t)$ is positive and bounded away from zero, uniformly on bounded subsets of $[\bar{t},\infty)$. The assertion of the theorem now follows by virtue of Theorem 4.1.

Definition 4.1 A characteristic $\eta(\cdot)$ on $[\bar{t},\infty)$ is called a shock if $u(\eta(t)-,t) > u(\eta(t)+,t)$ for all $t \in (\bar{t},\infty)$.

Definition 4.2 A point $(\bar{x},\bar{t}) \in (-\infty,\infty) \times (0,\infty)$ is called a shock generation point if the (unique) forward characteristic through $(\bar{x},\bar{t})$ is a shock while every backward characteristic through $(\bar{x},\bar{t})$ is genuine on $[0,\bar{t}]$.

Definition 4.3 A point $(\bar{x},\bar{t}) \in (-\infty,\infty) \times (0,\infty)$ is called the center of a centered compression wave if there are two genuine backward characteristics $\zeta_1(\cdot)$ and $\zeta_2(\cdot)$ through $(\bar{x},\bar{t})$ and every backward characteristic through $(\bar{x},\bar{t})$ contained in the funnel confined between $\zeta_1(\cdot)$ and $\zeta_2(\cdot)$(*) is genuine on $[0,\bar{t}]$.

(*) By the theory of contingent equations of type (3.1) [15], the funnel confined between $\zeta_1(\cdot)$ and $\zeta_2(\cdot)$ will be completely filled by backward characteristics.

It is clear that if $(\bar{x},\bar{t})$ is a shock generation point, then either $u(\bar{x}-,\bar{t}) = u(\bar{x}+,\bar{t})$, so that $\zeta_-(t;\bar{x},\bar{t}) \equiv \zeta_+(t;\bar{x},\bar{t})$ and there is just one backward characteristic through $(\bar{x},\bar{t})$, or $u(\bar{x}-,\bar{t}) > u(\bar{x}+,\bar{t})$ in which case $(\bar{x},\bar{t})$ is the center of a centered compression wave confined between $\zeta_-(t;\bar{x},\bar{t})$ and $\zeta_+(t;\bar{x},\bar{t})$.

Our next project is to study the structure of shocks.

Lemma 4.1 Let $\eta(\cdot)$ be a shock defined on $[\bar{t},\infty)$, $\bar{t} > 0$. Consider the functions $w_-(t) \overset{\text{def}}{=} u(\eta(t)-,t)$, $w_+(t) \overset{\text{def}}{=} u(\eta(t)+,t)$. Then $w_\pm(t)$ are continuous from the right on $[\bar{t},\infty)$. Furthermore, limits $w_\pm(t-)$ from the left exist for all $t \in (\bar{t},\infty)$ and $w_-(t-) \leqslant w_-(t)$, $w_+(t-) \geqslant w_+(t)$.

Proof. By Corollary 3.2, $\sigma > \tau \geqslant \bar{t}$ implies $\zeta_-(s;\eta(\sigma),\sigma) < \zeta_-(s;\eta(\tau),\tau)$, $s \in (0,\tau]$. Therefore, for fixed $t \geqslant \bar{t}$,

$$\zeta_-(s;\eta(\tau),\tau) \longrightarrow \zeta_-(s;\eta(t),t), \quad s \in [0,t], \text{ as } \tau \longrightarrow t+. \tag{4.5}$$

Similarly, for $t > \bar{t}$,

$$\zeta_-(s;\eta(\tau),\tau) \longrightarrow \zeta_o(s), \quad s \in [0,t), \text{ as } \tau \longrightarrow t-, \tag{4.6}$$

where $\zeta_o(\cdot)$ is a backward characteristic through $(\eta(t),t)$.

From Theorem 4.1 and (4.5) we conclude that, as $\tau \longrightarrow t+$, the slope $f'(w_-(\tau))$ of $\zeta_-(s;\eta(\tau),\tau)$ converges to the slope $f'(w_-(t))$ of $\zeta_-(s;\eta(t),t)$ so that $w_-(t+)$ exists and equals $w_-(t)$. Similarly, from (4.6) we deduce that, as $\tau \longrightarrow t-$, the slope $f'(w_-(\tau))$ of $\zeta_-(s;\eta(\tau),\tau)$ converges to the slope $\dot{\zeta}_o$ of $\zeta_o(\cdot)$ which cannot exceed the slope of $\zeta_-(s;\eta(t),t)$. Hence $w_-(t-)$ exists and is less than or equal to $w_-(t)$. The analogous properties of $w_+(\cdot)$ are established by the same procedure. The proof is complete.

In particular, the set of points of discontinuity of $w_{\pm}(\cdot)$ is at most countable. We note that the characteristic $\zeta_o(\cdot)$ constructed in the proof of Lemma 4.1 is genuine. Indeed, by (4.6) and Theorem 4.1 we deduce

$$u(\zeta_o(s)+,s) = \lim_{\tau \to t-} u(\zeta_-(s;\eta(\tau),\tau) \pm,s) = w(t-), \quad s \in (0,t).$$

Recalling that $\dot{\zeta}_o(s) = f'(w(t-))$ and applying Theorem 4.1 we obtain $u(\zeta_o(s)+,s) = u(\zeta_o(s)-,s) = w(t-)$, $s \in (0,t)$. If t is a point of continuity of $w_-(\cdot)$, $\zeta_o(s)$ coincides with $\zeta_-(s;\eta(t),t)$. If, however, $w_-(t-) < w_-(t)$, then $\zeta_-(s;\eta(t),t) < \zeta_o(s)$, $s \in [0,t)$. In this case the funnel confined between $\zeta_-(s;\eta(t),t)$ and $\zeta_o(s)$ is filled with backward characteristics through $(\eta(t),t)$. If all these characteristics are genuine, $(\eta(t),t)$ is the center of a centered compression wave; otherwise $(\eta(t),t)$ is a point of shocks interaction. Analogous arguments apply to $w_+(\cdot)$. These considerations together with Theorem 3.1 and Lemma 4.1 yield

<u>Theorem 4.3</u> Let $\eta(\cdot)$ be a shock on $[\bar{t},\infty)$, $\bar{t} > 0$. Then the derivative of $\eta(\cdot)$ from the right exists at every $t \in [\bar{t},\infty)$ and is given by

$$D^+\eta(t) = \frac{f(u(\eta(t)+,t))-f(u(\eta(t)-,t))}{u(\eta(t)+,t)-u(\eta(t)-,t)}. \tag{4.7}$$

Furthermore, $\dot{\eta}(t)$ exists and is continuous except on the (at most countable) set of points of interaction of $\eta(\cdot)$ with another shock and/or a centered compression wave. At every point of interaction the strength $u(\eta(t)-,t)-u(\eta(t)+,t)$ of the shock increases.

In particular, any shock may interact with an at most countable set of other shocks. This observation combined with Corollary 4.1 yields

<u>Corollary 4.2</u> The set of shocks is at most countable.

We now investigate the smoothness of $u(x,t)$, jointly in the (x,t) variables.

Theorem 4.4 Let $(\bar{x},\bar{t}) \in (-\infty,\infty) \times (0,\infty)$ and let $\eta(\cdot)$ be the (unique) forward characteristic through $(\bar{x},\bar{t})$. Then $(\bar{x},\bar{t})$ is a point of continuity of $u(x,t)$ relative to the set $\mathscr{S}_- = \{(x,t) | t \geq \bar{t},\ x < \eta(t) \text{ or } t < \bar{t},\ x \leq \zeta_-(t;\bar{x},\bar{t})\}$, the limit being $u(\bar{x}-,\bar{t})$, and also relative to the set $\mathscr{S}_+ = \{(x,t) | t \geq \bar{t},\ x > \eta(t) \text{ or } t < \bar{t},\ x \geq \zeta_+(t;\bar{x},\bar{t})\}$ with limit $u(\bar{x}+,\bar{t})$.

Proof. Every sequence in $\mathscr{S}_-$ converging to $(\bar{x},\bar{t})$ must contain a subsequence $\{(x_n,t_n)\}$ such that $\{\zeta_-(t;x_n,t_n)\}$ converges to a backward characteristic through $(\bar{x},\bar{t})$. Since $\zeta_-(t;x,s) \leq \zeta_-(t;\bar{x},\bar{t})$, for any $(x,s) \in \mathscr{S}_-$, it follows that $\zeta_-(t;x_n,t_n) \longrightarrow \zeta_-(t;\bar{x},\bar{t})$, $t \in (0,\bar{t})$. Thus the slope $f'(u(x_n-,t_n))$ of $\zeta_-(t;x_n,t_n)$ converges to the slope $f'(u(\bar{x}-,\bar{t}))$ of $\zeta_-(t;\bar{x},\bar{t})$ and this shows that $(\bar{x},\bar{t})$ is a point of continuity of $u(x,t)$ relative to $\mathscr{S}_-$ with limit $u(\bar{x}-,\bar{t})$. Similarly one shows that $(\bar{x},\bar{t})$ is also a point of continuity relative to $\mathscr{S}_+$ and the limit is $u(\bar{x}+,\bar{t})$. The proof is complete.

Corollary 4.3 The solution $u(x,t)$ is continuous at the point $(\bar{x},\bar{t}) \in (-\infty,\infty) \times (0,\infty)$ if and only if $u(\bar{x}-,\bar{t}) = u(\bar{x}+,\bar{t})$.

Theorem 4.5 Let $\eta(\cdot)$ be a shock and let $\bar{t}$ be a point of continuity of the functions $w_{\pm}(t) = u(\eta(t)\pm,t)$. Then $(\eta(\bar{t}),\bar{t})$ is a point of continuity of $u(x,t)$ relative to the set $\mathscr{S}_- = \{(x,t) | x < \eta(t)\}$ the limit being $u(\eta(\bar{t})-,\bar{t})$, and also relative to the set $\mathscr{S}_+ = \{(x,t) | x > \eta(t)\}$ with limit $u(\eta(\bar{t})+,\bar{t})$.

Proof. Any sequence in $\mathscr{S}_-$ converging to $(\eta(\bar{t}),\bar{t})$ must contain a subsequence $\{(x_n,t_n)\}$ such that $\{\zeta_-(t;x_n,t_n)\}$ converges to a backward characteristic through $(\eta(\bar{t}),\bar{t})$. By Corollary 3.2, $\zeta_-(t;x_n,t_n) < \zeta_-(t;\eta(t_n),t_n)$, $t \in (0,t_n]$. Moreover, (see the discussion following the proof of Lemma 4.1) since $\bar{t}$ is a point of continuity of $w_-(\cdot)$, $\zeta_-(t;\eta(t_n),t_n) \longrightarrow \zeta_-(t;\eta(\bar{t}),\bar{t})$, $t \in (0,\bar{t})$. Therefore, $\zeta_-(t;x_n,t_n) \longrightarrow \zeta_-(t;\eta(\bar{t}),\bar{t})$, $t \in (0,\bar{t})$. In particular, the slope $f'(u(x_n-,t_n))$ converges to the slope $f'(u(\eta(\bar{t})-,\bar{t}))$ of $\zeta_-(t;\eta(\bar{t}),\bar{t})$ and this implies that $(\eta(\bar{t}),\bar{t})$ is a point of continuity relative to $\mathscr{S}_-$ with limit $u(\eta(\bar{t})-,\bar{t})$. Similarly one shows that $(\eta(\bar{t}),\bar{t})$ is also a point of continuity relative to $\mathscr{S}_+$ and the limit is $u(\eta(\bar{t})+,\bar{t})$. The proof is complete.

From the above results the following picture of the structure of solutions emerges: There is an at most countable set of shocks. With the exception of interaction points, limits of the solution exist on both sides of a shock and the Rankine-Hugoniot condition is satisfied. On the complement of the shock set the solution is continuous.

Focussing of classical characteristics is the cause of generation of shocks. Spreading of characteristics has of course the opposite effect, namely of smoothing the solution. Therefore, it should be expected that the solution is smoother in the interior of the set of points of continuity, where there is room for the spreading of characteristics. Such a result was indeed established by DiPerna [10] in the (considerably more difficult) case of genuinely nonlinear systems endowed with Riemann invariants. We present here a similar result for (1.1).

Theorem 4.6 If $u(x,t)$ is continuous on a neighborhood of a point $(\bar{x},\bar{t}) \in (-\infty,\infty) \times (0,\infty)$, then $f'(u(x,t))$ is Lipschitz continuous at $(\bar{x},\bar{t})$. Furthermore, if $f''(u(\bar{x},\bar{t})) > 0$, then $u(x,t)$ is Lipschitz continuous at $(\bar{x},\bar{t})$.

Proof. Assume that $u(x,t)$ is continuous on the circle $B_r = \{(x,t) | |x-\bar{x}|^2 + |t-\bar{t}|^2 \leqslant r^2\}$, $r > 0$. Take any pont (x,t) at a distance $\rho < r$ from $(\bar{x},\bar{t})$. The characteristics through (x,t) and $(\bar{x},\bar{t})$ are genuine (and thus straight lines) inside B_r and cannot intersect in B_r (unless, of course, they coincide). Some elementary trigonometric considerations then show that the difference $|f'(u(x,t))-f'(u(\bar{x},\bar{t}))|$ of the slopes of these characteristics cannot exceed $c\rho/r$ where $c = 1 + \max\{[f'(u(y,s))]^2 | (y,s) \in B_r\}$. It follows that $f'(u(x,t))$ is Lipschitz continuous at $(\bar{x},\bar{t})$.

If $f''(u(\bar{x},\bar{t})) > 0$, for r sufficiently small $f''(u(y,s)) \geqslant a > 0$ if $(y,s) \in B_r$. Therefore, $|u(x,t) - u(\bar{x},\bar{t})| \leqslant \frac{c\rho}{ar}$ and $u(x,t)$ is Lipschitz continuous at $(\bar{x},\bar{t})$. The proof is complete.

Nothing precludes of course the possibility that the shock set is everywhere dense in the upper half-plane, in which case the interior of the set of points of continuity is empty.

Glimm and Lax [12] have observed that the spreading of characteristics is the mechanism that causes the decrease, in time, of the variation of solutions. This phenomenon manifests itself in the following proposition that establishes an estimate discovered by Hopf [1] (in the case $f(u) = \frac{1}{2}u^2$) and Oleinik (e.g. [3]).

<u>Theorem 4.7</u> For $-\infty < x < y < \infty$ and $t > 0$,

$$\frac{f'(u(y\pm,t)) - f'(u(x\pm,t))}{y - x} \leqslant \frac{1}{t}. \tag{4.8}$$

In particular, the increasing (and consequently also the total) variation of $f'(u(\cdot,t))$ is locally bounded on $(-\infty,\infty)$. Furthermore, if $f''(u) \geqslant a > 0$ for u in the range of the solution,

$$\frac{u(y\pm,t) - u(x\pm,t)}{y - x} \leqslant \frac{1}{at} \tag{4.9}$$

so that $u(\cdot,t)$ is of locally bounded variation on $(-\infty,\infty)$.

Proof. By Corollary 3.2, $\zeta_\pm(0;x,t) \leqslant \zeta_\pm(0;y,t)$. On account of Theorem 4.1, $\zeta_\pm(0;x,t) = x-tf'(u(x\pm,t))$, $\zeta_\pm(0;y,t) = y-tf'(u(y\pm,t))$ and (4.8) follows. If $f''(u) \geqslant a > 0$, (4.8) and the mean-value theorem yield (4.9). The proof is complete.

In the next section we study smoothness of solutions when the initial data are smooth.

5 SMOOTHNESS OF SOLUTIONS

In this section we assume that $f(\cdot)$ is C^{k+1} smooth, $3 \leqslant k \leqslant \infty$, and we study the smoothness of admissible solutions $u(x,t)$ of (1.1) with C^k smooth initial data

$$u(x,0) = w(x), \quad x \in (-\infty,\infty). \tag{5.1}$$

For $(x,t) \in (-\infty,\infty) \times (0,\infty)$ we consider the interceptors $\zeta_-(0;x,t)$ and $\zeta_+(0;x,t)$ of the backward minimal and maximal characteristic through (x,t). By Theorem 4.1,

$$x = y + t\, f'(w(y)) \tag{5.2}$$

$$u(x\pm,t) = w(y) \tag{5.3}$$

where

$$y = \zeta_\pm(0;x,t). \tag{5.4}$$

By the arguments employed repeatedly in Section 4, it is clear that, for fixed $\bar{t} \in (0,\infty)$, $\zeta_-(0;x,\bar{t})$ and $\zeta_+(0;x,\bar{t})$ are strictly increasing functions of x, continuous from the left and right, respectively. In particular, for any $(\bar{x},\bar{t}) \in (-\infty,\infty) \times (0,\infty)$,

$$1 + \bar{t}\, \frac{d}{dy}\, f'(w(y)) \geqslant 0, \; y = \zeta_\pm\, (0;\bar{x},\bar{t}). \tag{5.5}$$

Every point $(\bar{x},\bar{t})$ of continuity of $u(x,t)$ is also a point of continuity of $\zeta_\pm\,(0;x,t)$. (Of course at a point of continuity $\zeta_-(0;\bar{x},\bar{t}) = \zeta_+(0{:}\bar{x},\bar{t})$.) From this observation, (5.2), (5.3), (5.4) and the implicit function theorem we obtain

<u>Lemma 5.1</u> If $(\bar{x},\bar{t}) \in (-\infty,\infty) \times (0,\infty)$ is a point of continuity of $u(x,t)$ and

$$1 + \bar{t}\,\frac{d}{dy}\,f'(w(y)) > 0, \quad y = \zeta_{\pm}\,(0;\bar{x},\bar{t}), \tag{5.6}$$

then $u(x,t)$ is C^k smooth on a neighborhood of $(\bar{x},\bar{t})$.

From (5.5) and Lemma 5.1 we deduce the following corollary.

<u>Lemma 5.2</u> If $(\bar{x},\bar{t})$ is a point of continuity of $u(x,t)$ as well as a shock generation point, then

$$1 + \bar{t}\,\frac{d}{dy}\,f'(w(y)) = 0, \quad y = \zeta_{\pm}\,(0;\bar{x},\bar{t}). \tag{5.7}$$

We now consider a point $(\bar{x},\bar{t})$ of discontinuity of $u(x,t)$. If $(\bar{x},\bar{t})$ lies on a shock $\eta(\cdot)$ and $\bar{t}$ is a point of continuity of $u(\eta(t)\pm,t)$, then, using Theorem 4.5, (5.2), (5.3), (5.4) and the implicit function theorem we conclude that, if (5.6) is satisfied, there are C^k smooth functions $u_-(x,t)$ and $u_+(x,t)$, defined on some neighborhood $\mathfrak{N}$ of $(\bar{x},\bar{t})$, such that $u(x,t) = u_-(x,t)$ on $\{(x,t) \in \mathfrak{N} | x < \eta(t)\}$ and $u(x,t) = u_+(x,t)$ on $\{(x,t) \in \mathfrak{N} | x > \eta(t)\}$. This combined with Theorem 4.3 yields

<u>Lemma 5.3</u> If $(\bar{x},\bar{t})$ is a point on a shock $\eta(\cdot)$, (5.6) is satisfied, and $\bar{t}$ is a point of continuity of $u(\eta(t)\pm,t)$, then $\eta(\cdot)$ is C^{k+1} smooth on a neighborhood of $\bar{t}$ and $u(x,t)$ is C^k smooth on either side of $\eta(\cdot)$, on some neighborhood of $(\bar{x},\bar{t})$.

Finally we consider the case where $(\bar{x},\bar{t})$ is the center of a centered compression wave. By Definition 4.3,

<u>Lemma 5.4</u> If $(\bar{x},\bar{t})$ is the center of a centered compression wave confined

between the genuine characteristics $\zeta_1(\cdot)$ and $\zeta_2(\cdot)$, then

$$y + \bar{t}\, f'(w(y)) = \bar{x} \quad \text{for all } y \in [\zeta_1(0), \zeta_2(0)]. \tag{5.8}$$

In particular,

$$1 + \bar{t}\, \frac{d}{dy} f'(w(y)) = 0 \quad \text{for all } y \in [\zeta_1(0), \zeta_2(0)]. \tag{5.9}$$

We now observe that, for fixed y, $1 + tdf'(w(y))/dy$ as a function of t, changes sign across its zeros. In combination with (5.5) this implies that if for certain $(\bar{x},\bar{t})$ (5.5) is satisfied as an equality, then the forward characteristic through $(\bar{x},\bar{t})$ is necessarily a shock, that is $(\bar{x},\bar{t})$ is a point of the shock set. It follows that on the complement of the shock set (5.5) is satisfied as a strict inequality in which case Lemma 5.1 yields the following corollary.

Theorem 5.1 The complement of the shock set is open and $u(x,t)$ is C^k smooth on this set.

In preparation of the discussion of generic smoothness of solutions we now prove

Lemma 5.5 If $(\bar{x},\bar{t}) \in (-\infty,\infty) \times (0,\infty)$ is a point of continuity of $u(x,t)$ as well as a shock generation point, then $\zeta_\pm(0;\bar{x},\bar{t})$ is a critical point of the function $\frac{d}{dy} f'(w(y))$.

Proof. Let $\eta(\cdot)$ denote the shock generated at $(\bar{x},\bar{t})$. We consider a strictly decreasing sequence $\{t_n\}$, $t_n \longrightarrow \bar{t}+$, and we set $y_n^- = \zeta_-(0;\eta(t_n),t_n)$, $y_n^+ = \zeta_+(0;\eta(t_n),t_n)$, $\bar{y} = \zeta_\pm(0;\eta(\bar{t}),\bar{t})$. On account of Lemma 4.1, $y_n^- \longrightarrow \bar{y}-$, $y_n^+ \longrightarrow \bar{y}+$. By Lemma 5.2,

$$\frac{d}{dy} f'(w(y)) = -\frac{1}{\bar{t}}, \; y = \bar{y}.$$

On the other hand, using (5.5),

$$\frac{d}{dy} f'(w(y)) \geq -\frac{1}{t_n} > -\frac{1}{\bar{t}}, \ y = y_n^{\pm}, \ n = 1,2, \ldots .$$

Therefore, $\bar{y}$ is a critical point of $\frac{d}{dy} f'(w(y))$. The proof is complete.

Similarly, by Lemma 5.4,

Lemma 5.6 If $(\bar{x},\bar{t})$ is the center of a centered compression wave confined between the genuine characteristics $\zeta_1(\cdot)$ and $\zeta_2(\cdot)$, then every point of the interval $[\zeta_1(0), \zeta_2(0)]$ is a critical point of the function $\frac{d}{dy} f'(w(y))$.

We will say that the solution $u(x,t)$ is piecewise smooth if every bounded subset of $(-\infty,\infty) \times [0,\infty)$ intersects an at most finite number of shocks, every shock is a piecewise continuously differentiable curve, and $u(x,t)$ is C^k smooth on the complement of the shock set.

The following proposition exhibits a property of initial data that generate solutions which are not piecewise smooth.

Lemma 5.7 Suppose that $u(x,t)$ is not piecewise smooth. Then there is $\bar{y} \in (-\infty,\infty)$ such that

$$f''(w(\bar{y})) > 0 \tag{5.10}$$

and

$$\frac{d^m}{dy^m} f'(w(y)) = 0, \ m = 2,\ldots, k, \ y = \bar{y}. \tag{5.11}$$

Proof. Since $u(x,t)$ is not piecewise smooth, there are numbers $T > 0$ and a,b, $-\infty < a < b < \infty$, such that the set $S = \{(x,t) | 0 \leq t < T, \ \zeta_-(t;a,T) < x < \zeta_+(t;b,T)\}$ has the following property: S is intersected by an infinite set of shocks and/or at least one of the shocks intersecting S is not piecewise C^1 smooth inside S.

We note that the number of shocks intersecting S equals the number of shock generation points contained in S. Furthermore, by Theorem 4.3, differentiability of a shock may fail only at shock interaction points and/or at centers of centered compression waves. Therefore, using Lemmas 5.6 and 5.7, we conclude that $\frac{d}{dy} f'(w(y))$ must have an infinite number of critical points inside the interval $(\zeta_-(0;a,T), \zeta_+(0;b,T))$, corresponding to shock generation points and/or centers of centered compression waves. From this set of critical points we select a convergent sequence $\{y_n\}$ with limit, say, $\bar{y}$.

By virtue of Lemmas 5.2 and 5.4,

$$\frac{d}{dy} f'(w(y)) < -\frac{1}{T}, \quad y = y_n, \; n = 1,2, \ldots .$$

Consequently,

$$\frac{d}{dy} f'(w(y)) \leqslant -\frac{1}{T}, \quad y = \bar{y},$$

and this implies (5.10). Moreover, since $\bar{y}$ is a cluster point of the set of critical points of the C^{k-1} smooth function $\frac{d}{dy} f'(w(y))$, it follows that derivatives of this function of order 1, ..., k - 1 must vanish at $\bar{y}$, i.e., (5.11) is satisfied. The proof is complete.

We now have the preparation to show that solutions are generically piecewise smooth. Several variants of this result can be established, depending on the choice of topology. We will work here in the space C^k of functions that are bounded on $(-\infty,\infty)$ together with their derivatives of order 1,...,k, equipped with the topology induced by uniform convergence on $(-\infty,\infty)$ of derivatives of order 0,1,...,k.

Theorem 5.2 Generically, solutions of (1.1) generated by initial data in C^k are piecewise smooth and do not contain centered compression waves.

Proof. In view of Lemmas 5.6, 5.7 we have to show that the set of $w(\cdot)$ in C^k that satisfy (5.10), (5.11) for some $\bar{y} \in (-\infty,\infty)$ is of the first category in C^k. To this end it suffices to prove that for any fixed interval $[a,b]$ and any fixed number $\delta > 0$ the set of functions $w(\cdot)$ for which

$$f''(w(\bar{y})) \geqslant \delta, \tag{5.12}$$

$$\frac{d^m}{dy^m} f'(w(y)) = 0,\ m = 2,3,\quad y = \bar{y}, \tag{5.13}$$

for some $\bar{y} \in [a,b]$, is closed and nowhere dense in C^k.

This set is indeed closed because if $\{w_n(\cdot)\}$ is any sequence of functions in C^k for which $f''(w_n(y_n)) \geqslant \delta$, $\frac{d^m}{dy^m} f'(w_n(y)) = 0$, $m = 2,3$, $y = y_n$, for some $y_n \in [a,b]$, and $w_n(\cdot) \xrightarrow{C^k} w(\cdot)$, then it is clear that (5.12), (5.13) will be satisfied for $w(\cdot)$ with $\bar{y}$ any cluster point of the sequence $\{y_n\}$.

In order to show that the above set is also nowhere dense, we fix some $w(\cdot)$ in C^k and proceed to show that there are functions in C^k arbitrarily near $w(\cdot)$, for which (5.12), (5.13) are not jointly satisfied for any $\bar{y} \in [a,b]$.

The set of y with $f''(w(y)) > \delta/2$ is an open covering of the compact set $\{y \in [a,b] \mid f''(w(y)) \geqslant \delta\}$. Consequently, there is a finite subcovering. We can thus find numbers $a \leqslant a_1 < b_1 < \ldots < a_n < b_n \leqslant b$ with the following properties: $f''(y) \geqslant \delta/2$ for $y \in [a_i,b_i]$, $i = 1,\ldots, n$; $f''(a_i) = \delta/2$, $i = 1,\ldots,n$, unless $i = 1$ and $a_1 = a$; $f''(b_i) = \delta/2$, $i = 1,\ldots,n$, unless $i = n$ and $b_n = b$; the set of $y \in [a,b]$ with $f''(y) \geqslant \delta$ is contained in $\bigcup_{i=1}^{n} [a_i,b_i]$.

With each interval $[a_i,b_i]$ we associate $\varepsilon_i > 0$ satisfying the following conditions: $\varepsilon_i < (a_i - b_{i-1})/2$, $i = 2,\ldots,n$; $f''(y) \geqslant \delta/4$ if $y \in [a_i - \varepsilon_i, a_i]$. We construct C^∞ smooth functions $\phi_i(y)$ with the following properties:

$\phi_i(\cdot)$ is near zero in C^∞; the support of $\phi_i(\cdot)$ is contained in the interval $(a_i-\varepsilon_i,\infty)$; all critical points (if any) of the function $\frac{d}{dy} f'(w(y)) + \phi_i(y)$ on the interval $[a_i,b_i]$ are nondegenerate.(*)

We now construct a function $v(\cdot)$ in C^k as follows: We define $v(y) = w(y)$ for $y \leqslant a_1 - \varepsilon_1$. On $[a_i-\varepsilon_i,b_i]$ $v(\cdot)$ is defined as the solution of the initial value problem

$$\frac{d}{dy} f'(v(y)) = \frac{d}{dy} f'(w(y)) + \phi_i(y) \quad .$$
$$v(a_i - \varepsilon_i) = w(a_i - \varepsilon_i). \tag{5.14}$$

Finally, on $(b_i,a_{i+1} - \varepsilon_{i+1})$, $i = 1,\dots, n - 1$, and on (b_n,∞) we define $v(y)$ in such a way that $v(\cdot)$ is in C^k and is near $w(\cdot)$ in C^k.

We note that this construction of $v(\cdot)$ is possible because, since $f''(w(y)) \geqslant \delta/4$, for $y \in [a_i - \varepsilon_i, b_i]$,we will have, for $\phi_i(\cdot)$ sufficiently near zero, a well-posed problem in (5.14) and the resulting solution $v(\cdot)$ will be near $w(\cdot)$ on $[a_i - \varepsilon_i, b_i]$. We also observe that, for $v(\cdot)$ sufficiently near $w(\cdot)$, the set of $y \in [a,b]$ with $f''(v(y)) \geqslant \delta$ is contained in $\bigcup_{i=1}^{n} [a_i,b_i]$, because $f''(w(y)) < \delta$ on the compact interval $[b_i,a_{i+1}]$. On the other hand, by virtue of $(5.14)_1$ and the construction of $\phi_i(\cdot)$, all critical points (if any) of $\frac{d}{dy} f'(v(y))$ on $[a_i,b_i]$ are nondegenerate. We have thus constructed functions arbitrarily near $w(\cdot)$ for which (5.12),

(*) It is well-known that the set A of values of $\frac{d^2}{dy^2} f'(w(y))$ at points where $\frac{d^3}{dy^3} f'(w(y)) = 0$ has Lebesgue measure zero. Hence, if we arrange so that $\phi_i(y) = -ky, y \in [a_i,b_i]$, where $k \notin A$, we have $\frac{d}{dy}[\frac{d}{dy} f'(w(y))+\phi_i(y)] \neq 0$ if $\frac{d^2}{dy^2}[\frac{d}{dy} f'(w(y))+\phi_i(y)] = 0$, $y \in [a_i,b_i]$.

(5.13) are not jointly satisfied for any $\bar{y} \in [a,b]$. Therefore, the set of $w(\cdot)$ in C^k for which (5.12), (5.13) are satisfied for some $\bar{y} \in [a,b]$ is nowhere dense in C^k. The proof is complete.

6 A USEFUL IDENTITY

We establish here an identity satisfied by pairs of admissible solutions and explore some of its implications.

Theorem 6.1 Let $u(x,t)$, $\bar{u}(x,t)$ be two admissible solutions of (1.1). Then, for any fixed $x,\bar{x} \in (-\infty,\infty)$ and $t \in (0,\infty)$,

$$\int_{\bar{x}}^{x}[u(y,t)-\bar{u}(y,t)]\,dy - \int_{\bar{x}-tf'(\bar{u}(\bar{x}\pm,t))}^{x-tf'(u(x\pm,t))}[u(y,0)-\bar{u}(y,0)]\,dy$$

$$= -\int_{0}^{t}\{f(u(x\pm,t)) - f(\bar{u}(x+(\tau-t)f'(u(x\pm,t))\pm,\tau))$$

$$- f'(u(x\pm,t))[u(x\pm,t) - \bar{u}(x+(\tau-t)f'(u(x\pm,t))\pm,\tau)]\}d\tau$$

$$-\int_{0}^{t}\{f(\bar{u}(\bar{x}\pm,t)) - f(u(\bar{x}+(\tau-t)f'(\bar{u}(\bar{x}\pm,t))\pm,\tau))$$

$$- f'(\bar{u}(\bar{x}\pm,t))[\bar{u}(\bar{x}\pm,t) - u(\bar{x}+(\tau-t)f'(\bar{u}(\bar{x}\pm,t))\pm,\tau)]\}d\tau. \qquad (6.1)$$

Proof. It is an immediate corollary of Lemma 3.2 and Theorem 4.1.

The usefulness of identity (6.1) lies in that both integrals on the right-hand side are non-positive, in view of the convexity of $f(\cdot)$. Thus, if we have control on the second integral on the left-hand side of (6.1), we may use this identitiy to derive bounds on the first integral on the left-hand side as well as on either integral on the right-hand side. Using this approach we reestablish in this section certain known a priori estimates for solutions of (1.1) and, in subsequent sections, we obtain results on the asymptotic behavior of solutions.

Corollary 6.1 Let $u(x,t)$ $\bar{u}(x,t)$ be two admissible solutions of (1.1) with $u(x,0) \geqslant \bar{u}(x,0)$, a.e. on $(-\infty,\infty)$. Then $u(x \pm ,t) \geqslant \bar{u}(x \pm ,t)$ for all $x \in (-\infty,\infty)$, $t \in (0,\infty)$.

Proof. We need only observe that if, for some $t \in (0,\infty)$ and an interval $[\bar{x},x]$, $u(y-,t) < \bar{u}(y-,t)$, then $\bar{x}-tf'(\bar{u}(\bar{x}-,t)) < \bar{x}-tf'(u(\bar{x}-,t)) \leqslant x-tf'(u(x-,t))$ so that (6.1) yields a contradiction.

Corollary 6.2 Let $v(x,t)$, $w(x,t)$ be two admissible solutions of (1.1) with ranges contained in the interval $[m,M]$. Then, for any $-\infty < \bar{x} < x < \infty$ and $t > 0$,

$$\int_{\bar{x}}^{x} |v(y,t) - w(y,t)|dy \leqslant \int_{\bar{x}+tf'(m)}^{x+tf'(M)} |v(y,0) - w(y,0)|dy. \tag{6.2}$$

Proof. We decompose the interval $[\bar{x},x]$ into an (at most countable) set of disjoint intervals on which $v(y,t) - w(y,t)$ does not change sign and we apply (6.1) separately for each one of the above intervals with $u(y,t) \overset{\text{def}}{=} v(y,t)$, $\bar{u}(y,t) \overset{\text{def}}{=} w(y,t)$ if $v(y,t) - w(y,t) \leqslant 0$ and $u(y,t) \overset{\text{def}}{=} w(y,t)$, $\bar{u}(y,t) \overset{\text{def}}{=} v(y,t)$ if $v(y,t) - w(y,t) \geqslant 0$. Summing up the partial inequalities, we arrive at (6.2). The proof is complete.

For alternative proofs of Corollaries 6.1 and 6.2 under more general assumptions (several space variables, no convexity hypothesis on $f(\cdot)$) see, e.g., [7].

The next proposition, which gives necessary and sufficient conditions for a characteristic to remain genuine for all $t \geqslant 0$, will find several applications in subsequent sections.

Lemma 6.1 Let $u(x,t)$ be an admissible solution of (1.1). The straight line $\xi(t) = z + tf'(\bar{u})$, with z, $\bar{u}$ constants, is a genuine characteristic for all $t \geqslant 0$ if and only if

$$\int_z^{\chi} [u(y,0) - \bar{u}]\,dy \geqslant 0 \quad \text{for all } \chi \in (-\infty,\infty). \tag{6.3}$$

Proof. Assume first (6.3) is satisfied. We fix $t > 0$ and apply (6.1) with $\bar{x} = x = \xi(t) = z + tf'(\bar{u})$ to get

$$\int_z^{x-tf'(u(x\pm,t))} [u(y,0) - \bar{u}]\,dy \leqslant 0$$

which, on account of (6.3), must hold as an equality. But then both integrals on the right-hand side of (6.1) must vanish in the present case and this yields $u(x\pm,t) = \bar{u}$. Consequently, $\xi(\cdot)$ is a genuine characteristic for all $t \geqslant 0$.

Conversely, assume $\xi(\cdot)$ is a genuine characteristic for all $t \geqslant 0$. We fix $\chi \in (-\infty,\infty)$ and let $\bar{v}$ be a constant such that $\bar{v} > \bar{u}$ if $\chi < z$, $\bar{v} < \bar{u}$ if $\chi > z$. Consider the solution $\bar{u}(x,t) \overset{\text{def}}{=} \bar{v}$ and the characteristic $\eta(t) = \chi + tf'(\bar{v})$ which intersects $\xi(\cdot)$ at a point, say, $(\xi(t),t)$. Applying (6.1) with $\bar{x} = x = \xi(t)$ we obtain

$$-\int_{\chi}^{z} [u(y,0) - \bar{v}]\,dy \geqslant 0.$$

Letting $\bar{v} \longrightarrow \bar{u}$ we arrive at (6.3). The proof is complete.

As a corollary of the above proposition we reestablish a result of Lax [2] on the existence of invariants of solutions of (1.1).

Theorem 6.2 Let $u(x,t)$ be an admissible solution of (1.1) such that

$\int_{-\infty}^{z}[u(y,0) - \bar{u}]\,dy$ exists for some constant $\bar{u}$. Then $\int_{-\infty}^{z}[u(y,t) - \bar{u}]\,dy$ exists for all $t \geqslant 0$ and

$$\inf_x \int_{-\infty}^{x}[u(y,t) - \bar{u}]\,dy = \inf_z \int_{-\infty}^{z}[u(y,0) - \bar{u}]\,dy \overset{def}{=} I_- . \tag{6.4}$$

Proof. Applying (6.1), with $\bar{x} = -\infty$, we immediately deduce that $\int_{-\infty}^{x}[u(y,t) - \bar{u}]\,dy$ exists and

$$\inf_x \int_{-\infty}^{x}[u(y,t) - \bar{u}]\,dy \geqslant \inf_z \int_{-\infty}^{z}[u(y,0) - \bar{u}]\,dy = I_- . \tag{6.5}$$

On the other hand, if $I_- = \inf_z \int_{-\infty}^{z}[u(y,0)-\bar{u}]\,dy$ is realized as a minimum at a point $z \in (-\infty,\infty)$, $\int_{z}^{\chi}[u(y,0)-\bar{u}]\,dy \geqslant 0$ for all $\chi \in (-\infty,\infty)$ so that, by Lemma 6.1, $\xi(t) = z + tf'(\bar{u})$ is a genuine characteristic for all $t \geqslant 0$. Thus, applying (6.1) with $\bar{x} = -\infty$, $x = \xi(t)$, we obtain

$$\int_{-\infty}^{\xi(t)} [u(y,t)-\bar{u}]\,dy = I_- . \tag{6.6}$$

Combining (6.5) with (6.6) we arrive at (6.4). The cases where I_- is realized at $z = -\infty$ or $z = +\infty$ are treated similarly. The proof is complete.

We close this section with a derivation of an interesting characterization of admissible solutions due to Lax [2]. We define the function

$$g(v) = \max_u\, [uv - f(u)] \tag{6.7}$$

noting that the maximum is attained at $u = [f']^{-1}(v)$.

<u>Theorem 6.3</u> Let $u(x,t)$ be an admissible solution of (1.1). Consider the function

$$G(x,t,y) \overset{\text{def}}{=} \int_0^y u(z,0)dz + tg\left(\frac{x-y}{t}\right), \tag{6.8}$$

defined for $x \in (-\infty,\infty)$, $t \in (0,\infty)$, $y \in (-\infty,\infty)$. For fixed (x,t) $G(x,t,y)$ attains its minimum at a point $\bar{y}$ if and only if the straight line joining (x,t) with $(\bar{y},0)$ is a genuine characteristic.

<u>Proof</u>. We fix $y,\bar{y} \in (-\infty,\infty)$ and apply Lemma 3.2 to obtain

$$\begin{aligned}\int_y^{\bar{y}} u(z,0)dz &+ \int_0^t [f(u(y + \tau\,\frac{x-y}{t} \pm,\ \tau)) - \frac{x-y}{t}\,u(y + \tau\,\frac{x-y}{t} \pm,\ \tau)]d\tau \\ &- \int_0^t [f(u(\bar{y} + \tau\,\frac{x-\bar{y}}{t} \pm,\ \tau)) - \frac{x-\bar{y}}{t}\,u(\bar{y} + \tau\,\frac{x-\bar{y}}{t} \pm,\ \tau)]d\tau = 0.\end{aligned} \tag{6.9}$$

Now if the straight line joining (x,t) with $(\bar{y},0)$ is a genuine characteristic with slope, say, $f'(\bar{u}) = \frac{x-\bar{y}}{t}$, then

$$f(u(\bar{y}+\tau\,\frac{x-\bar{y}}{t} \pm,\tau)) - \frac{x-\bar{y}}{t}\,u(\bar{y}+\tau\,\frac{x-\bar{y}}{t} \pm,\tau) = f(\bar{u}) - \bar{u}\,f'(\bar{u}) = -\,g\left(\frac{x-\bar{y}}{t}\right)$$

$$f(u(y+\tau\,\frac{x-y}{t} \pm,\tau)) - \frac{x-y}{t}\,u(y+\tau\,\frac{x-y}{t} \pm,\tau) \geqslant \min_u\,[f(u) - \frac{x-y}{t}u] = -\,g\left(\frac{x-y}{t}\right)$$

so that (6.9) yields

$$G(x,t,\bar{y}) \leqslant G(x,t,y).$$

Conversely, if y is a minimum of $G(x,t,\cdot)$, we choose $\bar{y}$ so that the straight line joining(x,t) with $(\bar{y},0)$ is a genuine characteristic (e.g., we choose $\bar{y} = \zeta_-\ (0;x,t))$ and retracing our steps we observe that $G(x,t,y) = G(x,t,\bar{y})$ only if $f'(u(y + \tau\,\frac{x-y}{t} \pm,\tau)) = \frac{x-y}{t}$ which shows that the straight line joining (x,t) with $(y,0)$ is indeed a genuine characteristic. The proof is complete.

7 DECAY OF SOLUTIONS

In this section we employ (6.1) to establish decay of solutions of (1.1), as $t \to \infty$. It turns out that the rate of decay depends on two factors, namely the rate of growth of the primitive of the initial data and the 'strictness' of the convexity of $f(u)$ at $u = 0$.

Theorem 7.1 Assume that for u in a neighborhood of 0,

$$c|u|^p \leqslant f''(u) \leqslant C|u|^p, \quad p \geqslant 0, \; 0 < c \leqslant C. \tag{7.1}$$

Let $u(x,t)$ be an admissible solution of (1.1) with

$$\int_x^{x+L} u(y,0)dy = O(L^s), \; s \in [0,1), \text{ as } L \to \infty, \tag{7.2a}$$

or

$$\int_x^{x+L} u(y,0)dy = o(L^s), \; s \in [0,1], \text{ as } L \to \infty, \tag{7.2b}$$

uniformly in x on $(-\infty,\infty)$. Then

$$u(x\pm,t) = O(t^{\frac{s-1}{p(1-s)+2-s}}), \text{ as } t \to \infty, \tag{7.3a}$$

or

$$u(x\pm,t) = o(t^{\frac{s-1}{p(1-s)+2-s}}), \text{ as } t \to \infty, \tag{7.3b}$$

uniformly in x on $(-\infty,\infty)$.

Proof. We apply (6.1) with $\bar{u}(x,t) \equiv 0$ and $\bar{x} = x$ thus obtaining

$$-\int_{x-tf'(0)}^{x-tf'(u(x\pm,t))} u(y,0)dy \geqslant -t\,[f(u(x\pm,t)) - f(0) - f'(u(x\pm,t))\,u(x\pm,t)]. \quad (7.4)$$

Using (7.2) we deduce from (7.4)

$$\Phi(u(x\pm,t)) = O(t^{s-1}), \text{ as } t \to \infty, \quad (7.5a)$$

or

$$\Phi(u(x\pm,t)) = o(t^{s-1}), \text{ as } t \to \infty, \quad (7.5b)$$

uniformly in x on $(-\infty,\infty)$, where

$$\Phi(u) \overset{\text{def}}{=} \frac{-f(u)+f(0)+uf'(u)}{|f'(u)-f'(0)|^s} = \frac{\int_0^u vf''(v)dv}{|\int_0^u f''(v)dv|^s}. \quad (7.6)$$

A simple computation shows that $\Phi(0) = 0$, $\Phi'(u) > 0$ for $u > 0$, $\Phi'(u) < 0$ for $u < 0$. Moreover, on account of (7.1), (7.6), and for u near 0,

$$\Phi(u) \geqslant K|u|^{p(1-s)+2-s} \quad (7.7)$$

and this, combined with (7.5), yields (7.3). The proof is complete.

Important special cases of Theorem 7.1 were considered earlier by several authors. In the case (7.2b) with s = 1, corresponding to initial data with zero mean value, (7.3b) implies that the solution decays to zero uniformly on $(-\infty,\infty)$. This result was established by Lax [2]. Similarly, the case (7.2a) with s = 0, corresponding to initial data with bounded primitive (e.g., data in $L^1(-\infty,\infty)$, periodic data with zero mean value etc.) was discussed by Greenberg and Tong [17] and, for $p = 0$, by Lax [2]. More detailed information on the asymptotic behavior of solutions is available in certain special cases (data with compact support, periodic data) which will be discussed in Sections 9 and 10.

8 INITIAL CONDITIONS ASYMPTOTIC TO RIEMANN DATA. THE SHOCK CASE

We assume here that $f(\cdot)$ is locally uniformly convex, i.e., $f''(u) > 0$ for all $u \in (-\infty,\infty)$, and we study the asymptotic behavior of an admissible solution $u(x,t)$ with initial conditions $u(x,0)$ such that the improper integrals

$$\int_{-\infty}^{0}[u(x,0)-u_-]\,dx \quad , \quad \int_{0}^{\infty}[u(x,0)-u_+]\,dx \tag{8.1}$$

exist where u_- and u_+ are constants with

$$u_- > u_+ \, . \tag{8.2}$$

We select the origin $x = 0$ so that

$$\int_{-\infty}^{0}[u(x,0) - u_-]\,dx + \int_{0}^{\infty}[u(x,0) - u_+]\,dx = 0 \tag{8.3}$$

and proceed to compare $u(x,t)$ with the function

$$\bar{u}(x,t) = \begin{cases} u_- & x < K\,t \\ u_+ & x > K\,t \end{cases} \tag{8.4}$$

$$K = \frac{f(u_+)-f(u_-)}{u_+-u_-} \tag{8.5}$$

which is the solution of (1.1) with initial conditions

$$\bar{u}(x,0) = \begin{cases} u_- & x < 0 \\ u_+ & x > 0 \, . \end{cases} \tag{8.6}$$

Ilin and Oleinik [19,20], using the viscosity method, show that, for any $\delta > 0$, $u(x,t) \longrightarrow \bar{u}(x,t)$, as $t \longrightarrow \infty$, uniformly in x on $(-\infty,Kt-\delta)$ and $(Kt+\delta,\infty)$.

By using (6.1) we establish here the following more precise result.

Theorem 8.1 Let $\eta(\cdot)$ be any forward characteristic through (0,0). Then, as $t \longrightarrow \infty$,

$$\eta(t) = Kt + o(1) \tag{8.7}$$

$$u(x\pm,t) = \begin{cases} u_- + o(t^{-\frac{1}{2}}) & \text{uniformly for } x < \eta(t) \\ u_+ + o(t^{-\frac{1}{2}}) & \text{uniformly for } x > \eta(t) \end{cases} \tag{8.8}$$

Proof. We first observe that, as $t \longrightarrow \infty$,

$$\zeta_-(0;\eta(t),t) \longrightarrow -\infty, \qquad \zeta_+(0;\eta(t),t) \longrightarrow \infty. \tag{8.9}$$

Indeed, if this were not the case, $\zeta_-(0;\eta(t),t)$ and/or $\zeta_+(0;\eta(t),t)$ would converge, as $t \longrightarrow \infty$, to genuine characteristics defined for all $t > 0$ (asymptotes of $\eta(\cdot)$)which is impossible since, in view of (8.1) and (8.2), (6.3) cannot be satisfied for any $\bar{u} \in (-\infty,\infty)$.

We now fix $(x,t) \in (-\infty,\infty) \times (0,\infty)$ and we apply (6.1) with $\bar{u}(x,t)$ given by (8.4) and $\bar{x} = x$. On account of (8.1), (8.3) and (8.9), the left-hand side of (6.1) is $o(1)$ as $t \longrightarrow \infty$, uniformly in x on $(-\infty,\infty)$. Thus each term on the right-hand side of (6.1) must be $o(1)$ as $t \longrightarrow \infty$. We distinguish the following cases:

(i) $x < \min\{\eta(t),Kt\}$. (8.10)

Then

$$t\{-f(u(x\pm,t))+f(u_-)+f'(u(x\pm,t))[u(x\pm,t)-u_-]\} = o(1) \tag{8.11}$$

and this yields (8.8) for the case (8.10) by virtue of the uniform convexity of $f(\cdot)$.

(ii) $\quad x > \max\{\eta(t), Kt\}$. (8.12)

Then

$$t\{-f(u(x\pm,t))+f(u_+)+f'(u(x\pm,t))[u(x\pm,t)-u_+]\} = o(1) \tag{8.13}$$

which gives (8.8) for the case (8.12).

(iii) $\quad Kt \leqslant x < \eta(t)$. (8.14)

Then

$$(t-\tau_\pm)\{-f(u(x\pm,t))+f(u_+)+f'(u(x\pm,t))[u(x\pm,t)-u_+]\}$$

$$+\ \tau_\pm\{-f(u(x\pm,t))+f(u_-)+f'(u(x\pm,t))[u(x\pm,t)-u_-]\} = o(1) \tag{8.15}$$

with $\tau_\pm$ determined by $\zeta_\pm(\tau_\pm, x, t) = K\tau_\pm$. Now it is clear that $\tau_\pm \longrightarrow \infty$ as $t \longrightarrow \infty$ and this implies $u(x\pm,t) = u_- + o(1)$. Then (8.15) yields $t - \tau_\pm = o(1)$ as $t \longrightarrow \infty$, independently of the choice of x. This result implies (8.7), whenever $Kt < \eta(t)$, and also gives

$$t\{-f(u(x\pm,t))+f(u_-)+f'(u(x\pm,t))[u(x\pm,t)-u_-]\} = o(1) \tag{8.16}$$

which yields (8.8) for the case (8.14).

(iv) $\quad \eta(t) < x \leqslant Kt$. (8.17)

Then

$$(t-\tau_\pm)\{-f(u(x\pm,t))+f(u_-)+f'(u(x\pm,t))[u(x\pm,t)-u_-]\}$$

$$+\tau_\pm\{-f(u(x\pm,t))+f(u_+)+f'(u(x\pm,t))[u(x\pm,t)-u_+]\} = o(1) \tag{8.18}$$

with $\tau_\pm$ determined by $\zeta_\pm(\tau_\pm ; x,t) = K\tau_\pm$. As above, $\tau_\pm \longrightarrow \infty$, as $t \longrightarrow \infty$, and gives $u(x\pm,t) = u_+ + o(1)$. Then (8.18) implies $t - \tau_\pm = o(1)$, as $t \longrightarrow \infty$, independently of the choice of x. We thus obtain (8.7), whenever $\eta(t) < Kt$, and also

$$t\{-f(u(x\pm,t))+f(u_+)+f'(u(x\pm,t))[u(x\pm,t)-u_+]\} = o(1). \qquad (8.19)$$

Finally, (8.19) yields (8.8) in the case (8.17) and thus completes the proof.

Corollary 8.1 In the special case where $u(x,0) = u_+$ for $x > b$ and $u(x,0) = u_-$ for $x < a$, $u(x,t) = \bar{u}(x,t)$ for t sufficiently large and all $x \in (-\infty,\infty)$.

9 INITIAL CONDITIONS ASYMPTOTIC TO RIEMANN DATA. THE WAVE CASE

In this section we assume that f is C^3 smooth and $f''(u) > 0$ for all $u \in (-\infty,\infty)$ and we study the asymptotic behavior of an admissible solution $u(x,t)$ with initial conditions $u(x,0)$ satisfying

$$u(x,0) = \begin{cases} u_- & x < -L \\ u_+ & x > L \end{cases} \tag{9.1}$$

where $L > 0$ and u_-, u_+ are constants such that

$$u_- \leqslant u_+ \ . \tag{9.2}$$

We shall compare $u(x,t)$ with the function

$$\bar{u}(x,t) = \begin{cases} u_- & x < tf'(u_-) \\ [f']^{-1}\left(\frac{x}{t}\right) & tf'(u_-) \leqslant x \leqslant tf'(u_+) \\ u_+ & tf'(u_+) < x \end{cases} \tag{9.3}$$

which is the solution of (1.1) with initial conditions

$$\bar{u}(x,0) = \begin{cases} u_- & x < 0 \\ u_+ & x > 0 \ . \end{cases} \tag{9.4}$$

Theorem 9.1 As $t \longrightarrow \infty$,

$$u(x,t) = \begin{cases} u_- & x < \eta_-(t) \\ u_- + \frac{1}{f''(u_-)}\left[\frac{x}{t} - f'(u_-)\right] + O\left(\frac{1}{t}\right) & \eta_-(t) < x < tf'(u_-) \\ [f']^{-1}\left(\frac{x}{t}\right) + O\left(\frac{1}{t}\right) & tf'(u_-) \leqslant x \leqslant tf'(u_+) \\ u_+ + \frac{1}{f''(u_+)}\left[\frac{x}{t} - f'(u_+)\right] + O\left(\frac{1}{t}\right) & tf'(u_+) < x < \eta_+(t) \\ u_+ & \eta_+(t) < x \end{cases} \tag{9.5}$$

where

$$\begin{aligned} \eta_-(t) &= t\, f'(u_-) - [2pt\, f''(u_-)]^{\frac{1}{2}} + O(1), \\ \eta_+(t) &= t\, f'(u_+) + [2qt\, f''(u_+)]^{\frac{1}{2}} + O(1), \end{aligned} \tag{9.6}$$

$$\begin{aligned} p &= -\min_y \int_{-\infty}^{y} [u(x,0) - u_-]dx, \\ q &= \max_y \int_{y}^{\infty} [u(x,0) - u_+]dx. \end{aligned} \tag{9.7}$$

Proof. Let $\eta_-(\cdot)$ and $\eta_+(\cdot)$ be, respectively, the minimal and maximal forward characteristic through $(-L,0)$ and $(L,0)$. It is clear that $u(x,t) = u_-$ for $x < \eta_-(t)$, $u(x,t) = u_+$ for $x > \eta_+(t)$. On account of (3.1), $\dot{\eta}_-(t) \leqslant f'(u_-)$ and $\dot{\eta}_+(t) \geqslant f'(u_+)$ so that $\eta_-(t) < tf'(u_-)$ and $\eta_+(t) > tf'(u_+)$.

Consider any x with $tf'(u_-) \leqslant x \leqslant tf'(u_+)$. Then $\zeta_\pm(0;x,t) \in [-L,L]$. Therefore,

$$f'(u(x\pm,t)) = \frac{x-\zeta_\pm(0;x,t)}{t} = \frac{x}{t} + O\left(\frac{1}{t}\right). \tag{9.8}$$

Since $f''(\cdot) > 0$, this implies (9.5), line 3.

We now fix x with $\eta_-(t) < x < tf'(u_-)$. Then $\zeta_\pm(0;x,t) \in [-L,L]$. Hence,

$$f'(u(x\pm,t)) = \frac{x-\zeta_\pm(0;x,t)}{t} = \frac{x}{t} + O\left(\frac{1}{t}\right) < f'(u_-) + O\left(\frac{1}{t}\right). \tag{9.9}$$

Consequently,

$$u(x\pm,t) \leqslant u_- + O\left(\frac{1}{t}\right). \tag{9.10}$$

We now apply (6.1) with $\bar{u}(x,t)$ given by (9.3) and $\bar{x} = x$ thus obtaining

$$\int_0^t \{f(u(x\pm,t)) - f(\bar{u}(\zeta_\pm(\tau;x,t)\pm,\tau)) \tag{9.11}$$

$$- f'(u(x\pm,t))[u(x\pm,t) - \bar{u}(\zeta_\pm(\tau;x,t)\pm,\tau)]\}d\tau = O(1).$$

Since $f''(\cdot) > 0$, we deduce from (9.11)

$$\int_0^t [u(x\pm,t) - \bar{u}(\zeta_\pm(\tau;x,t)\pm,\tau)]^2\, d\tau = O(1). \tag{9.12}$$

By virtue of (9.10) and noting that $\bar{u}(\zeta_\pm(\tau;x,t)\pm,\tau) \geqslant u_-$ we obtain from (9.12)

$$u(x\pm,t) = u_- + O(t^{-\frac{1}{2}}). \tag{9.13}$$

On account of (9.13) and since $f(\cdot)$ is C^3 smooth, (9.9) yields

$$f'(u_-) + f''(u_-)[u(x\pm,t) - u_-] + O\left(\frac{1}{t}\right) = \frac{x}{t} + O\left(\frac{1}{t}\right) \tag{9.14}$$

which gives (9.5), line 2.

By a similar procedure one establishes (9.5), line 4.

In order to prove $(9.6)_1$, we write the Rankine-Hugoniot condition (3.2) and use (9.13), (9.14) to obtain

$$\dot{\eta}_-(t) = \frac{f(u(\eta_-(t)+,t))-f(u_-)}{u(\eta_-(t)+,t)-u_-}$$

$$= f'(u_-) + \tfrac{1}{2}f''(u_-)[u(\eta_-(t)+,t)-u_-] + 0\left(\frac{1}{t}\right) \tag{9.15}$$

$$= \frac{1}{2t}\,\eta_-(t) + \tfrac{1}{2}\,f'(u_-) + 0\left(\frac{1}{t}\right).$$

Integrating the above differential equation we get

$$\eta_-(t) = t\,f'(u_-) - \alpha\,t^{\frac{1}{2}} + 0(1). \tag{9.16}$$

In order to determine the constant α in (9.16), we make use of Theorem 6.2. From (9.5) it is easily seen that the function $\int_{-\infty}^{y}[u(x,t) - u_-]dx$ will asymptotically attain its minimum at $y = t\,f'(u_-)$. Using (9.5), (9.16), Theorem 6.2 and with p defined by $(9.7)_1$,

$$p = -\frac{1}{f''(u_-)}\left[\frac{y^2}{2t} - y\,f'(u_-)\right]_{t\,f'(u_-)-\alpha\,t^{\frac{1}{2}} + 0(1)}^{t\,f'(u_-)} + 0(t^{-\frac{1}{2}}) \tag{9.17}$$

whence, after a simple calculation, we deduce $\alpha = [2pf''(u_-)]^{\frac{1}{2}}$. Thus $(9.6)_1$ follows from (9.16).

One establishes $(9.6)_2$ by a similar procedure. The proof is complete.

Remark 9.1 In the case $u_- = u_+ = 0$ (initial data with compact support) (9.5) reduces to the classical N-wave profile [2]. For more detailed results in this case see Theorem 10.3.

Combining Theorem 9.1 with Corollary 6.1 we easily obtain the following result that was also established by Ilin and Oleinik [19,20] via the viscosity method.

Theorem 9.2 Let $u(x,t)$ be an admissible solution of (1.1) with initial conditions $u(x,0)$ satisfying

$$u(x,0) \longrightarrow \begin{cases} u_- \ , \text{ as } x \longrightarrow -\infty \\ u_+ \ , \text{ as } x \longrightarrow \infty \end{cases} \tag{9.18}$$

where u_- and u_+ are constants with $u_- \leqslant u_+$. Then, as $t \longrightarrow \infty$,

$$u(x,t) = \bar{u}(x,t) + o(1), \tag{9.19}$$

uniformly in x on $(-\infty,\infty)$, where $\bar{u}(x,t)$ denotes the solution (9.3) of (1.1) with initial conditions (9.4).

We discuss here certain cases where more detailed information on asymptotic behavior of solutions can be obtained. We will be assuming again that $f(\cdot)$ is C^3 smooth and $f''(u) > 0$ for all $u \in (-\infty,\infty)$.

Lemma 10.1 Let $u(x,t)$ be an admissible solution of (1.1) and suppose that $\eta_-(\cdot)$ and $\eta_+(\cdot)$ are two characteristics on $[0,\infty)$ with $\eta_-(t) < \eta_+(t)$, $t \in [0,\infty)$. Then, as $t \longrightarrow \infty$,

$$\underset{\eta_-(t) < x < \eta_+(t)}{\text{Decreasing Variation}}\ u(x,t) = O\left(\frac{1}{t}\right). \tag{10.1}$$

Proof. Consider any mesh $\eta_-(t) < y_1 \leqslant y_2 \leqslant \cdots < \eta_+(t)$ such that $u(y_{2k-1}-,t) > u(y_{2k}+,t)$, $k = 1, \cdots n$. Since

$$\eta_-(0) \leqslant \zeta_-(0;y_1,t) \leqslant \zeta_+(0;y_2,t) \leqslant \cdots \leqslant \zeta_-(0;y_{2n-1},t) \leqslant \zeta_+(0;y_{2n}+,t)$$
$$\leqslant \eta_+(0),$$

and

$$\frac{y_k - \zeta_\pm(0;y_k,t)}{t} = f'(u(y_k \pm,t)),$$

we obtain

$$\sum_{k=1}^{n} [f'(u(y_{2k-1}-,t)) - f'(u(y_{2k}+,t))] \leqslant \frac{\eta_+(0) - \eta_-(0)}{t}.$$

Hence the decreasing variation of $f'(u(\cdot,t))$ over $(\eta_-(t), \eta_+(t))$ is $O\left(\frac{1}{t}\right)$, as

$t \longrightarrow \infty$. Since $f''(\cdot) > 0$, we conclude that the decreasing variation of $u(\cdot,t)$ over $(\eta_-(t), \eta_+(t))$ is also $0\left(\frac{1}{t}\right)$, as $t \longrightarrow \infty$. The proof is complete.

Combining Lemma 10.1 with Theorem 9.1 we arrive at

Corollary 10.1 Under the assumptions and the notation of Theorem 9.1, the decreasing variation of $u(\cdot,t)$ on $(\eta_-(t), \eta_+(t))$ is $0\left(\frac{1}{t}\right)$, as $t \longrightarrow \infty$.

A second application of Lemma 10.1 is the following

Theorem 10.1 Let $u(x,t)$ be an admissible solution of (1.1) such that $u(x,0)$ (and thereby $u(x,t)$) is periodic in x. Then the variation of $u(\cdot,t)$ over a period is $0\left(\frac{1}{t}\right)$, as $t \longrightarrow \infty$.

Proof. Let w be the period and $\bar{u}$ the mean value of $u(x,0)$. The function

$$\int_0^y [u(x,0) - \bar{u}]\, dx \tag{10.2}$$

is continuous periodic of period w and therefore attains its minimum at points x_o+kw, $k = 0, \pm 1, \pm 2, \cdots$. By virtue of Lemma 6.1, the straight lines $\xi_k(t) = x_o+kw+tf'(\bar{u})$ are genuine characteristics on $[0,\infty)$. Applying Lemma 10.1 we thus conclude that the decreasing variation and hence also the total variation of $u(\cdot,t)$ over a period is $0\left(\frac{1}{t}\right)$, as $t \longrightarrow \infty$. The proof is complete.

In the periodic case it is possible to get more precise information on asymptotic behavior from more detailed information on initial data. To this end we need the following

Lemma 10.2 Let $u(x,t)$ be an admissible solution of (1.1) and assume that for some $\bar{u} \in (-\infty,\infty)$ the function (10.2) attains its minimum at points y_-,y_+, $y_- < y_+$, but at no point inside (y_-,y_+). Then, as $t \longrightarrow \infty$,

$$u(x\pm,t) = \begin{cases} \bar{u} + \dfrac{1}{f''(\bar{u})}\left[\dfrac{x-y_-}{t} - f'(\bar{u})\right] + o\left(\dfrac{1}{t}\right), & y_- + tf'(\bar{u}) \leqslant x < \eta(t) \\ \\ \bar{u} + \dfrac{1}{f''(\bar{u})}\left[\dfrac{x-y_+}{t} - f'(\bar{u})\right] + o\left(\dfrac{1}{t}\right), & \eta(t) < x \leqslant y_+ + tf'(\bar{u}) \end{cases} \tag{10.3}$$

where $\eta(\cdot)$ is a shock with

$$\eta(t) = \tfrac{1}{2}\,[y_- + y_+] + tf'(\bar{u}) + o(1). \tag{10.4}$$

Proof. By Lemma 6.1, $y_- + tf'(\bar{u})$ and $y_+ + tf'(\bar{u})$ are genuine characteristics on $[0,\infty)$. We note that, as $t \longrightarrow \infty$, the interceptors $\zeta_\pm(0;x,t)$, $y_- + tf'(\bar{u}) < x < y_+ + tf'(\bar{u})$, must accumulate at y_- or y_+. Indeed, if a sequence $\{\zeta_\pm(0;x_n,t_n)\}$ converged, as $t_n \longrightarrow \infty$, to $y_o \in (y_-,y_+)$, then $\{\zeta_\pm(t;x_n,t_n)\}$ should converge to a genuine characteristic $\xi(\cdot)$ on $[0,\infty)$ through $(y_o,0)$ in which case, by Lemma 6.1, y_o would be a minimum of the function (10.2), contrary to our hypothesis.

Thus, for large t there is $\eta(t)$ with the property that, if $y_- + tf'(\bar{u}) \leqslant x < \eta(t)$, then $\zeta_\pm(0;x,t)$ are near y_- while, if $\eta(t) < x \leqslant y_+ + tf'(\bar{u})$, then $\zeta_\pm(0;x,t)$ are near y_+. Hence

$$\begin{aligned} f'(u(x\pm,t)) &= \frac{x-\zeta_\pm(0;x,t)}{t} = \frac{x-y_-}{t} + o\left(\frac{1}{t}\right), \quad y_- + tf'(\bar{u}) \leqslant x < \eta(t) \\ f'(u(x\pm,t)) &= \frac{x-\zeta_\pm(0;x,t)}{t} = \frac{x-y_+}{t} + o\left(\frac{1}{t}\right), \quad \eta(t) < x \leqslant y_+ + tf'(\bar{u}). \end{aligned} \tag{10.5}$$

On the other hand, one deduces from Lemma 10.1 that, as $t \longrightarrow \infty$,

$$u(x\pm,t) = \bar{u} + O\left(\frac{1}{t}\right), \; y_- + tf'(\bar{u}) \leqslant x \leqslant y_+ + tf'(\bar{u}). \tag{10.6}$$

Combining (10.5) with (10.6) we easily obtain (10.3). It is also clear that, for large t, $\eta(t)$ is a shock.

In order to prove (10.4), we apply (6.1) with $\bar{u}(x,t) \equiv \bar{u}$, $\bar{x} = y_- + tf'(\bar{u})$, $x = y_+ + tf'(\bar{u})$ thus obtaining

$$\int_{y_-+tf'(\bar{u})}^{y_++tf'(\bar{u})} [u(y,t) - \bar{u}]\,dy = \int_{y_-}^{y_+} [u(y,0) - \bar{u}]\,dy = 0. \tag{10.7}$$

Substituting u(y,t) from (10.3) into (10.7) and after a rather lengthy computation we arrive at (10.4). The proof is complete.

Remark 10.1 Under the assumptions of Lemma 10.2, if u(x,0) is continuously differentiable and y_-, y_+ are nondegenerate minima of (10.2), i.e., $u_x(y_-,0) > 0$, $u_x(y_+,0) > 0$, then it is easy to see that, for large t, u(x,t) is strictly increasing and continuously differentiable on the intervals $(y_- + tf'(\bar{u}), \eta(t))$ and $(\eta(t), y_+ + tf'(\bar{u}))$. Thus $\eta(\cdot)$ is the only shock inside $[y_- + tf'(\bar{u}), y_+ + tf'(\bar{u})]$. Furthermore, since the slope of $\zeta_\pm(\tau;x,t)$ is $f'(\bar{u}) + O\left(\frac{1}{t}\right)$, $\zeta_\pm(0;x,t) = y_- + O\left(\frac{1}{t}\right)$, for $y_- + tf'(\bar{u}) \leqslant x < \eta(t)$, and $\zeta_\pm(0;x,t) = y_+ - O\left(\frac{1}{t}\right)$, for $\eta(t) < x \leqslant y_+ + tf'(\bar{u})$. Hence, the approximation in (10.5) and thereby also in (10.3) is improved from $o\left(\frac{1}{t}\right)$ to $O\left(\frac{1}{t^2}\right)$.

From Lemma 10.2 and Remark 10.1 we immediately deduce the following corollary.

Theorem 10.2 Let u(x,t) be an admissible solution of (1.1) with u(x,0) periodic. Then, as $t \longrightarrow \infty$, $u(\cdot,t)$ is asymptotic, to order $O\left(\frac{1}{t}\right)$, to a periodic sawtooth function. The number of "teeth" per period equals the number of points in a period interval on which (10.2) attains its minimum, $\bar{u}$ being the mean value of u(x,0). Furthermore, if u(x,0) is continuously differentiable and the minima of (10.2) nondegenerate, then, for large t, $u(\cdot,t)$ is strictly increasing and continuously differentiable between consecutive "teeth" and is approximated by the sawtooth function to the order $O(t^{-2})$.

The generic case of Theorem 10.2 where (10.2) has a unique minimum per period has been discussed by several authors.

Finally, we consider the asymptotic behavior of solutions with initial data of compact support, thus obtaining a refinement of Theorem 9.1 in the case $u_- = u_+ = 0$.

Theorem 10.3 Let $u(x,t)$ be an admissible solution of (1.1) with initial data $u(x,0)$ that are continuously differentiable and have compact support on $(-\infty,\infty)$. Assume further that $\int_{-\infty}^{y} u(x,0)dx$ attains its minimum at a unique point $\bar{y}$ and that this minimum is nondegenerate, i.e., $u_x(\bar{y},0) > 0$. Then, for large t, the solution contains precisely two shocks $\eta_-(\cdot)$, $\eta_+(\cdot)$ that are C^2 smooth. Moreover, $u(x,t) = 0$ for $x < \eta_-(t)$ and $x > \eta_+(t)$ while on the interval $(\eta_-(t),\eta_+(t))$ $u(\cdot,t)$ is strictly increasing and continuously differentiable.

Proof. Let $\eta_-(\cdot)$ and $\eta_+(\cdot)$ be as in Theorem 9.1. We claim that, as $t \longrightarrow \infty$, $\zeta_+(0;\eta_-(t),t) \longrightarrow \bar{y}-$, $\zeta_-(0;\eta_+(t),t) \longrightarrow \bar{y}+$. Indeed, the function $\zeta_+(0;\eta_-(t),t)$ is increasing and bounded from above. If $\zeta_+(0;\eta_-(t),t)$ converged to $z \neq \bar{y}$, then $\zeta_+(\tau;\eta(t),t)$ should converge to a genuine characteristic through $(z,0)$ on $[0,\infty)$. By Lemma 6.1 this would mean that $\int_{-\infty}^{y} u(x,0)dx$ attains its minimum at z, contrary to our hypothesis. The proof that $\zeta_-(0;\eta_+(t),t) \longrightarrow \bar{y}+$ is similar.

Since $\zeta_\pm(0;x,t) \in [\zeta_+(0;\eta_-(t),t), \zeta_-(0;\eta_+(t),t)]$, for any $x \in (\eta_-(t),\eta_+(t))$, and $u(x,0)$ is strictly increasing on a neighborhood of $\bar{y}$, we conclude that for large t $u(\cdot,t)$ is strictly increasing and C^1 smooth on $(\eta_-(t),\eta_+(t))$. Moreover, since $u(\eta_-(t)+,t)$ and $u(\eta_+(t)-,t)$ are C^1 smooth, for large t, it follows by (4.7) that $\eta_-(\cdot)$ and $\eta_+(\cdot)$ are C^2 smooth. The proof is complete.

The above result was also proved by Schaeffer [13] who demonstrated that the assumptions of Theorem 10.3 are generically satisfied. In conjunction with Theorem 5.2 this result shows that generically solutions generated by smooth initial data with compact support contain a (globally) finite number of shocks.

References

1 E. Hopf, The partial differential equation $u_t+uu_x = \mu u_{xx}$. Comm. Pure Appl. Math. 3 (1950),201-230.

2 P. D. Lax, Hyperbolic systems of conservation laws II. Comm. Pure Appl. Math. 10 (1957),537-566.

3 O. A. Oleinik, Discontinuous solutions of non-linear differential equations. Usp. Mat. Nauk (N.S.) 12 (3) (1957), 3-73. English translation: A.M.S. Transl., Ser. 2, 26, 95-172.

4 E. Conway and J. A. Smoller, Global solutions of the Cauchy problem for quasi-linear first-order equations in several space variables. Comm. Pure Appl. Math. 19 (1966), 95-105.

5 A. I. Volpert, The space BV and quasilinear equations. Mat. Sbornik (N.S.) 73 (115) (1967), 255-302. English translation: Math. USSR-Sbornik 2 (1967), 225-267.

6 S. N. Kruzkov, First order quasilinear equations in several independent variables. Mat. Sbornik (N.S.) 81 (123) (1970), 228-255. English translation: Math. USSR-Sbornik 10 (1970),217-243.

7 M. G. Crandall, The semigroup approach to first order quasilinear equations in several space variables. Israel J. Math. 12 (1972), 108-132.

8 O. A. Oleinik, The Cauchy problem for nonlinear equations in a class of discontinuous functions. Dokl. Akad. Nauk SSSR 95 (1954), 451-454. English translation: A.M.S. Transl., Ser. 2, 42, 7-12.

9 O. A. Oleinik, The Cauchy problem for nonlinear differential equations of the first order with discontinuous initial conditions. Trudy Moskov. Mat. Obsc. 5 (1956), 433-454.

10 R. J. DiPerna, Singularities of solutions of nonlinear hyperbolic systems of conservation laws. Arch. Rat. Mech. Anal. 60 (1975), 75-100.

11 J. Glimm, Solutions in the large for nonlinear hyperbolic systems of equations. Comm. Pure Appl. Math. 18 (1965),697-715.

12 J. Glimm and P. D. Lax, Decay of solutions of systems of nonlinear hyperbolic conservation laws. Mem. A.M.S. 101 (1970).

13 D. G. Schaeffer, A regularity theorem for conservation laws. Advances in Math. 11 (1973), 368-386.

14 M. Golubitsky and D. G. Schaeffer, Stability of shock waves for a single conservation law. Advances in Math. 16 (1975), 65-71.

15 A. F. Filippov, Differential equations with discontinuous right-hand side. Mat. Sbornik (N.S.) 51 (93) (1960), 99-128. English translation: A.M.S. Transl., Ser. 2, 42, 199-231.

16 C. M. Dafermos, Generalized characteristics and the structure of solutions of hyperbolic conservation laws. (to appear in Indiana Univ. Math. J.).

17 J. M. Greenberg and D. M. Tong, Decay of periodic solutions of $\partial u/\partial t + \partial f(u)/\partial x = 0$. J. Math. Anal. Appl. 43 (1973), 56-71.

18 R. J. DiPerna, Decay and asysmptotic behavior of solutions to nonlinear hyperbolic systems of conservation laws. Ind. Univ. J. 24 (1975), 1047-1071.

19 A. M. Ilin and O. A. Oleinik, Behavior of the solutions of the Cauchy problem for certain quasilinear equations for unbounded increase of the time. Dokl. Akad. Nauk SSSR 120 (1958), 25-28. English translation: A.M.S. Transl., Ser. 2,42, 19-23.

20 A. M. Ilin and O. A. Oleinik, Asymptotic behavior of solutions of the Cauchy problem for some quasi-linear equations for large values of time. Mat. Sbornik (N.S.) 51 (93) (1960),191-216.

21 C. M. Dafermos, Quasilinear hyperbolic systems that result from conservation laws. Nonlinear Waves, Ch. III, pp. 82-102. S. Leibovich and A. R. Seebass, Eds. Cornell University Press, Ithaca NY 1974.

Professor C. M. Dafermos, Lefschetz Center for Dynamical Systems,
Brown University, Providence, Rhode Island 02912, U.S.A.

J K HALE

Generic bifurcation with applications

1 INTRODUCTION

Suppose X,Z are Banach spaces, Λ is an open set in a Banach space, $M:X\times\Lambda \longrightarrow Z$ is continuously differentiable and consider the equation

$$M(x,\lambda) = 0 \tag{1.1}$$

for $x\in X, \lambda\in\Lambda$. Sometimes we require more derivatives on M and it always will be assumed that M has as many continuous derivatives as necessary.

A solution of Equation (1.1) is a point $(x,\lambda)\in X\times\Lambda$ such that Equation (1.1) is satisfied. Let $S\subset X\times\Lambda$ denote the set of solutions of Equation (1.1) and, for any $\lambda\in\Lambda$, let $S_\lambda = \{x\in X:(x,\lambda)\in S\}$. Since λ is usually a physical parameter, it is natural to discuss the dependence of S_λ on λ. In this general setting, the set S could be almost any closed set. For example, S might be as depicted in Figure 1.

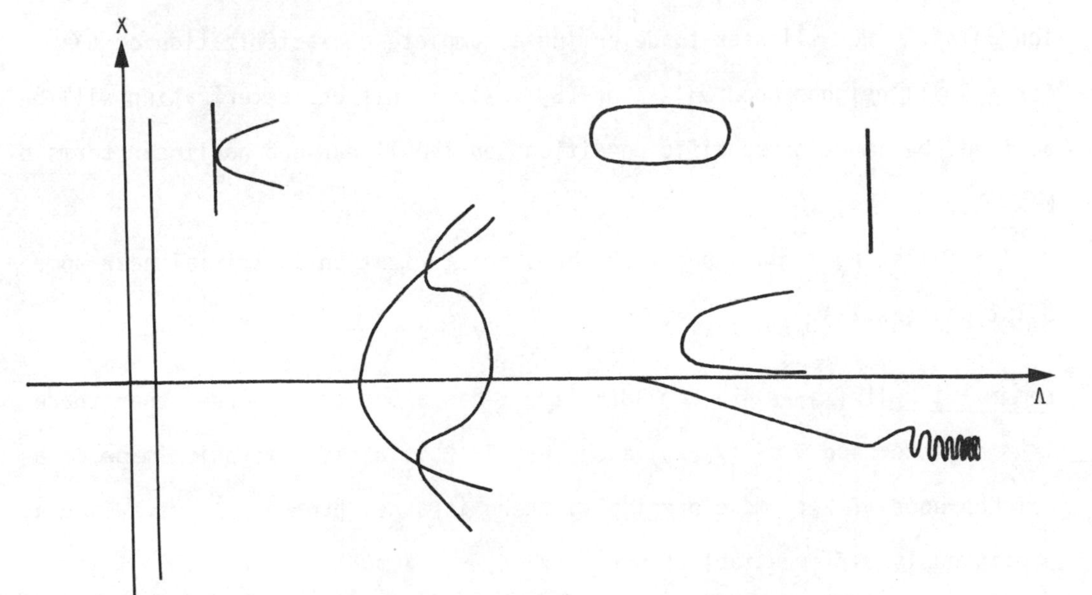

Figure 1

Thus, S_λ may go through drastic changes as λ varies. It can be an entire line or a linear subspace or a closed arc for a fixed λ_0 and be a single point for every λ close to but different from λ_0. At other values of λ_0, the number of solutions may be finite and this number may change by making small variations in λ. At some points λ_0, we can have either the same number or a different number of solutions at distinct points in an arbitrarily small neighborhood of λ_0. The solution set S need not even be represented as a continuous surface at λ_0.

The ultimate goal would be to describe completely the set S. In general, this is an impossible task. However, it is apparent that the most interesting points λ_0 are those for which the S_λ has different properties for points λ in an arbitrarily small neighborhood of λ_0. A <u>bifurcation point</u> λ_0 in Λ satisfies $S_{\lambda_0} \neq \phi$, the empty set, and there exists an $x_0 \in S_{\lambda_0}$ such that, for any neighborhood U of (x_0,λ_0), there are two distinct solutions (x_1,λ), $(x_2,\lambda) \in U$ and $x_1 \in S_\lambda$, $x_2 \in S_\lambda$. Our primary task will be to determine the nature of the solution set S_λ for λ in a neighborhood of a bifurcation point. We will seek to determine a complete characterization of $S \cap \Omega$ for a small neighborhood $\Omega \subseteq X \times \Lambda$ of $(S_{\lambda_0},\lambda_0)$. This characterization will be achieved by imposing specific conditions on the linear and nonlinear terms of $M(x_0,\lambda_0)$, $x_0 \in S_{\lambda_0}$.

The following lemma shows that the characterization is trivial near some solutions (x_0,λ_0).

<u>Lemma 1.1</u> If $(x_0,\lambda_0) \in S$ and $\partial M(x_0,\lambda_0)/\partial x$ has a bounded inverse, then there is a neighborhood Ω of (x_0,λ_0) such that $S \cap \Omega$ is a diffeomorphic image of a neighborhood of λ_0; more precisely, there is a neighborhood Λ_0 of λ_0 and a continuously differentiable function $x^*: \Lambda_0 \longrightarrow X$ such that $S \cap \Omega = \{(x^*(\lambda),\lambda), \lambda \in \Lambda_0\}$.

Proof. Since $M(x_0,\lambda_0)=0$ and $\partial M(x_0,\lambda_0)/\partial x$ has a bounded inverse, the implicit function theorem implies the result.

As a consequence of Lemma 1.1, the only points $(x_0,\lambda_0)\in S$ that require further discussion are those for which $\partial M(x_0,\lambda_0)/\partial x$ is singular. As remarked earlier, our approach for the discussion of these points is to impose conditions on $\partial M(x_0,\lambda_0)/\partial x$ and the other terms in $M(x_0,\lambda_0)$ which will lead to a complete characterization of S near (x_0,λ_0). To improve the intuition, let us discuss two elementary examples.

If λ, x are real scalars and

$$M(x,\lambda) \overset{\text{def}}{=} x^2-\lambda=0 \tag{1.2}$$

then the set S is given by the set $\{(x,\lambda):\lambda=x^2\}$. Obviously, $S_\lambda=\phi$, the empty set, if $\lambda<0$ (no solutions), $S_0=\{0\}$ (one solution), $S_\lambda=\{\pm\lambda^{\frac{1}{2}}\}$ if $\lambda>0$ (two solutions). The point $\lambda=0$ is a bifurcation point and the solutions lie on a nice smooth curve in the (x,λ)-plane (see Figure 2).

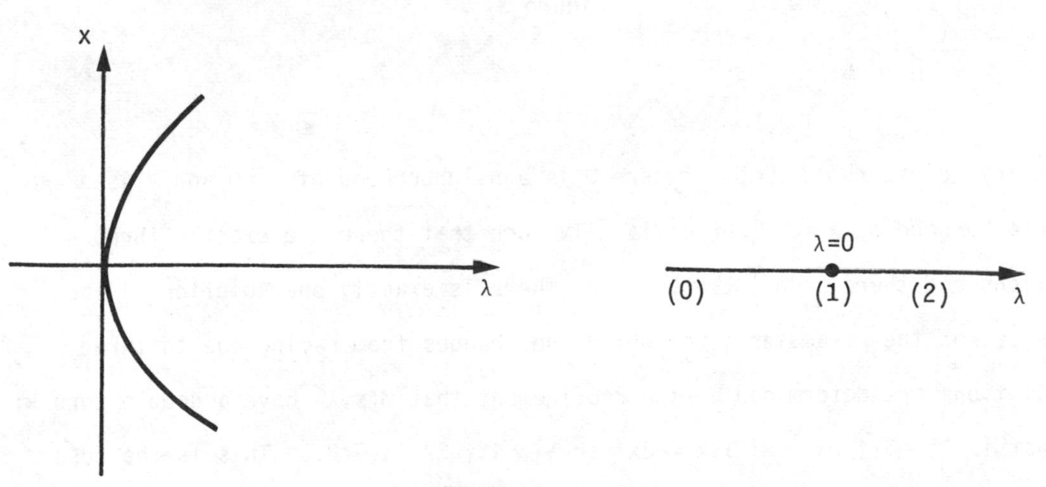

Figure 2

In the parameter space Λ, the number of solutions is labeled in parentheses.

As a second example, if x, λ_1, λ_2 are real scalars, $\lambda = (\lambda_1, \lambda_2)$ and

$$M(x,\lambda) \stackrel{\text{def}}{=} x^3 - \lambda_1 x - \lambda_2 = 0 \tag{1.3}$$

then the solution set S in $(x_1,\lambda_1\ \lambda_2)$-space is shown in Figure 3.

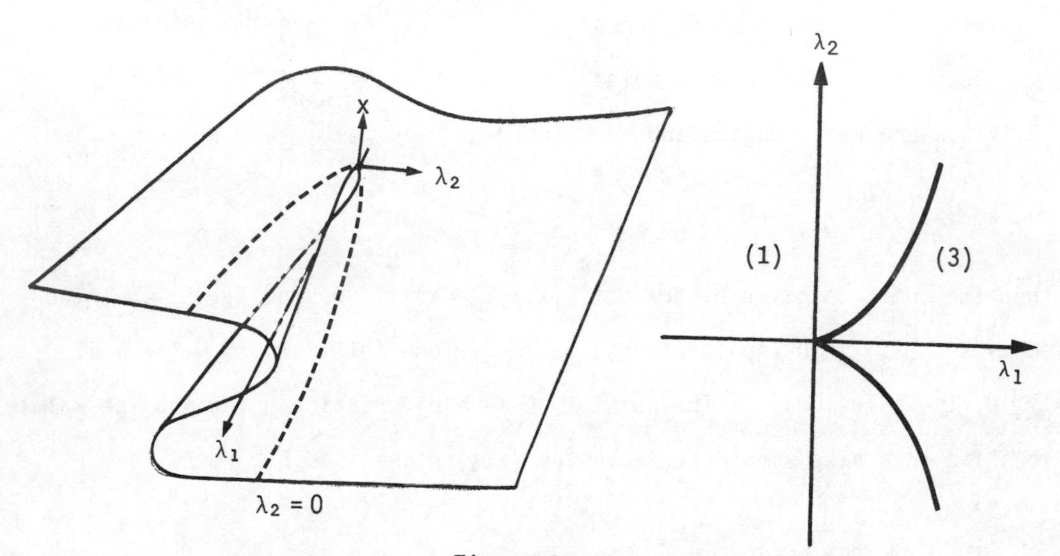

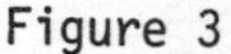
Figure 3

In any neighborhood $\Omega = U \times V$, where U is a neighborhood of $x = 0$ and V is a neighborhood of $\lambda = 0$, there is a $\bar{\lambda} \in V$ such that there are exactly three solutions and there is a $\bar{\bar{\lambda}} \in V$ such that there is exactly one solution. The values of the parameter λ for which one changes from having one to three solutions are determined by the requirement that $M(x,\lambda)$ have a double zero x; that is $\lambda_1 = 3x^2$, $\lambda_2 = x^3 - \lambda_1 x = -2x^3$ or $\lambda_2^2 = 4\lambda_1^3/27$, $\lambda_1 \geqslant 0$. This is the cusp shown in Figure 3. For ease in depicting the results, one also draws the picture x versus λ_1 for a fixed λ_2 as shown in Figure 4.

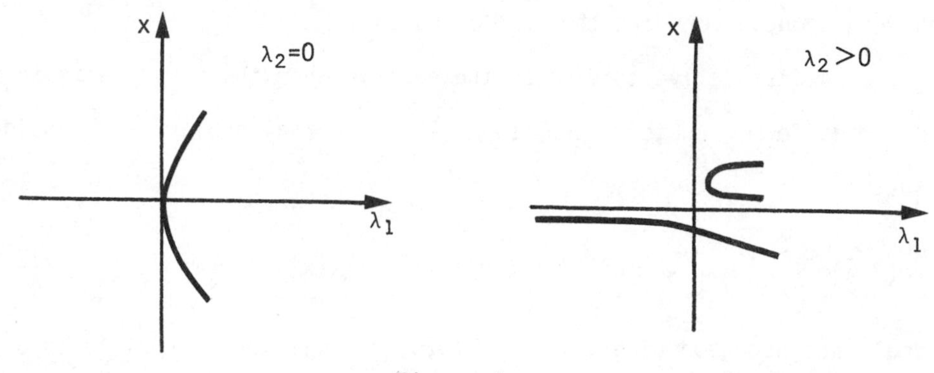

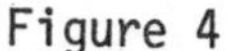

Figure 4

The cusp in Figure 3 is called the <u>bifurcation set</u> or <u>bifurcation curve</u> and represents the set in parameter space where the discriminant of the cubic polynomial in Equation (1.3) vanishes.

Although the two previous examples are very elementary, the method used in the discussion has more profound implications. As we will see in applications below, the functions in Equation (1.2), (1.3) are only the first terms in the Taylor series of more complicated functions. Let us consider this more general case. Suppose x,λ are real scalars and consider all solutions of the equation

$$M(x,\lambda) \overset{\text{def}}{=} x^2-\lambda + O(|x|^3 + |\lambda x| + |\lambda|^2) = 0 \tag{1.4}$$

in a small neighborhood of $(x,\lambda) = (0,0)$. Is the structure of the solution set going to be essentially the same as depicted in Figure 2? This is certainly the case as the following discussion shows. In a neighborhood of $(x,\lambda) = (0,0)$ there is a unique function $x^*(\lambda)$ such that $M(x,\lambda)$ is a minimum at $(x^*(\lambda),\lambda)$. If $\alpha(\lambda) \overset{\text{def}}{=} M(x^*(\lambda),\lambda)$, then $\alpha(\lambda) > 0$ implies there are no solutions of Equation (1.4), $\alpha(\lambda) = 0$ implies one and $\alpha(\lambda) < 0$ implies two.

It is obvious that $\alpha(\lambda) = -\lambda + O(|\lambda|^2)$ as $|\lambda| \longrightarrow 0$, $\alpha(\lambda)$ has the sign of $-\lambda$ and thus the conclusions are the same as before.

It is also instructive to discuss the example when the first terms in the Taylor series form a cubic. Suppose x, λ_1, λ_2 are real scalars and consider the equation

$$M(x,\lambda) \stackrel{\text{def}}{=} x^3 - \lambda_1 x - \lambda_2 + O(|x|^4 + |\lambda_1 x^2| + |\lambda_1^2 x| + |\lambda_2 x| + \lambda_2^2) = 0 \tag{1.5}$$

in a small neighborhood of $(x,\lambda_1,\lambda_2) = (0,0,0)$. If there is at least a double zero of $M(x,\lambda)$, then

$$\frac{\partial M(x,\lambda)}{\partial x} \stackrel{\text{def}}{=} 3x^2 - \lambda_1 + O(|x|^3 + |\lambda_1 x| + \lambda_1^2 + |\lambda_2|) = 0. \tag{1.6}$$

From the implicit function theorem, one can solve Equation (1.5),(1.6) uniquely for $\lambda_1^*(x)$, $\lambda_2^*(x)$ for x small and $\lambda_1 = 3x^2 + O(|x|^3)$, $\lambda_2 = -2x^3 + O(|x|^4)$ as $|x| \longrightarrow 0$. Therefore, $\lambda_1^3 = 27\lambda_2^2/4 + O(|\lambda_2|^3)$ as $|\lambda_2| \longrightarrow 0$. This is approximately the same cusp as in Figure 3. Along this cusp, $\partial^2 M(x,\lambda)/\partial x^2 \neq 0$ if $x \neq 0$ and $\partial M(x,\lambda)/\partial\lambda_2 \neq 0$. This is sufficient to conclude that the number of zeros of $M(x,\lambda)$ changes by exactly two as the cusp is crossed.

We now make some elementary but important remarks about these two examples. By imposing some conditions on the nonlinearities in the function (the square term in x is present in the first example and the cubic term in the second), we were able to determine the complete behavior of the solution set. Furthermore, we could determine the surface in parameter space (a point in the first example and a curve in the second) such that the number of solutions x of $M(x,\lambda) = 0$ changed as λ crossed this surface. Such surfaces are called <u>bifurcation surfaces</u> or <u>catastrophe surfaces</u>. Most of these notes are concerned with extension of these ideas to higher dimensions.

For the case where λ is a scalar, we do present some other types of results in Section 3.

In the literature, the term <u>branching of solutions</u> or <u>bifurcation of solutions</u> is very prevalent and we will continue to use it. If λ belongs to a bifurcation set, we say the solutions <u>bifurcate at λ</u> even though it does not begin to describe precisely the phenomena that occurs. The terminology arose in the following manner. If $M(0,\lambda) = 0$ for all λ, then the solution set $S \subset X \times \Lambda$ contains the set $S_0 = \{(0,\lambda), \lambda \in \Lambda\}$ and the problem is to determine all those solutions which branch or bifurcate from S_0. One can define the same concept of branching relative to any manifold of solutions in $X \times \Lambda$ provided it is locally diffeomorphic to a neighborhood of a point in Λ.

For the example, $M(x,\lambda) = x^2 - \lambda = 0$, there is no point on the curve in (x,λ)-space of solutions $x^2 = \lambda$ at which there is a branch point. On the other hand, we say the point $\lambda = 0$ in parameter space is a branch point or bifurcation point because the number of elements in $S_\lambda = \{x : x^2 = \lambda\}$ changes as λ crosses zero. In the example $M(x,\lambda) = x^3 - \lambda_1 x - \lambda_2 = 0$, if $\lambda_2 = 0$, the solution set does actually branch at $\lambda_1 = 0$ but does not if $\lambda_2 \neq 0$ although we will consistently use this terminology.

2 METHOD OF LIAPUNOV-SCHMIDT AND ALTERNATIVE PROBLEMS

In the local theory of bifurcation, one of the most important tools is the one commonly referred to as Liapunov-Schmidt, or more generally, alternative problems. For extensive references, see Bancroft, Hale and Sweet [5], Cesari [11,12], Gaines and Mawhin [25], Hale [31], Sather [72], Stakgold [78], Vainberg and Trenogin [87,88]. The basic idea of the method is to reduce the study of the solutions of Equation (1.1) which lie in a small neighborhood of a given point (x_0,λ_0) to a lower dimensional problem. The most typical case is to reduce the problem to one in which the unknown variable is an element of the null space of $\partial M(x_0,\lambda_0)/\partial x$. However, in some problems it is necessary to use an element from the generalized eigenspace of this operator and sometimes elements from even larger subspaces.

To begin, let us introduce some notation. The null space of a linear operator A is denoted by $\mathfrak{N}(A)$ and the range by $\mathcal{R}(A)$. If P is a continuous projection on a Banach space X, then X_P denotes $\mathcal{R}(P)$. By a translation of variables, we may always assume the point (x_0,λ_0) of interest is (0,0). Our objective is to determine all solutions of the equation

$$M(x,\lambda) = 0 \tag{2.1}$$

near (0,0). Basic to the investigation is the following hypothesis:

(H1) $M(0,0) = 0,\ \mathfrak{N}(B) = X_U,\ \mathcal{R}(B) = Z_E,\ B \overset{\text{def}}{=} \dfrac{\partial M(0,0)}{\partial x}.$

If (H1) is satisfied, then the closed graph theorem implies B has an inverse K on X_{I-U}. If $x = w + y$, $w = Ux$, $y = (I-U)x$, then Equation (2.1) is equivalent to

$$\begin{aligned} &(a) \quad EM(w+y,\lambda) = 0, \\ &(b) \quad (I-E)M(w+y,\lambda) = 0. \end{aligned} \tag{2.2}$$

If $F(w,y,\lambda) = EM(w+y,\lambda)$, then $F(0,0,0) = 0$ and $\partial F(0,0,0)/\partial y = E\partial M(0,0)/\partial x\colon X_{I-U} \longrightarrow Z$ has a bounded inverse. The implicit function theorem implies

<u>Lemma 2.1</u> There exist neighborhoods $Y \subseteq X_{I-U}$ of 0, $W \subseteq X_U$ of 0 and $\Lambda_0 \subseteq \Lambda$ of 0 such that Equation (2.2a) has a unique solution $y^*\colon W \times \Lambda_0 \longrightarrow Y$ which is continuous together with its first derivatives, $y^*(0,0) = 0$, $\partial y^*(0,0)/\partial w = 0$. Thus, the solutions $x = w + y$ of Equation (2.1) which lie in a sufficiently small neighborhood of 0 for $\lambda \in \Lambda_0$ must be given by $x = w + y^*(w,\lambda)$ where w satisfies the equation

$$0 = (I-E)M(w+y^*(w,\lambda),\lambda). \tag{2.3}$$

Equation (2.3) is usually referred to as the <u>bifurcation equation</u> or <u>determining equation</u>. For historical reasons, we adapt the first name even though these equations are not the equations for the bifurcation surfaces. The method of reduction to these equations is called the method of Liapunov-Schmidt or the method of alternative problems. Actually, the method of alternative problems is more general as has been pointed out clearly in the work of Cesari [11,12]. We need this more general method in only one situation; namely, where the undetermined element belongs to the generalized eigenspace of B. Thus, we repeat the above process under the following hypothesis:

(H2) There exist projections $\tilde{U}$ on X and $\tilde{E}$ on Z such that

$$BX_{I-\tilde{U}} \subseteq Z_{\tilde{E}},\ BX_{\tilde{U}} \subseteq Z_{I-\tilde{E}},\ X_U \subseteq X_{\tilde{U}}.$$

If $x = \tilde{w} + \tilde{y}, \tilde{w} = \tilde{U}x, \tilde{y} = (I-\tilde{U})x$, then $\tilde{y} \in X_{I-U}$ and B has the bounded inverse K on $X_{I-\tilde{U}}$. Since $\tilde{E}B(\tilde{w} + \tilde{y}) = B\tilde{y}$ and $(I-\tilde{E})B(\tilde{w} + \tilde{y}) = (I-\tilde{E})B\tilde{w}$, Equation (2.1) is equivalent to

$$\begin{aligned} &\text{(a)}\quad B\tilde{y} = \tilde{E}N(\tilde{w} + \tilde{y}, \lambda), \\ &\text{(b)}\quad (I-\tilde{E})[B\tilde{w} - N(\tilde{w} + \tilde{y}, \lambda)] = 0, \end{aligned} \tag{2.4}$$

where $N(x,\lambda) = Bx - M(x,\lambda)$, $N(0,0) = 0$, $\partial N(0,0)/\partial x = 0$.

As in the proof of Lemma 2.1, one obtains the following result.

Lemma 2.2 There exist neighborhoods $\tilde{Y} \subseteq X_{I-\tilde{U}}$ of 0, $\tilde{W} \subseteq X_{\tilde{U}}$ of 0 and $\Lambda_0 \subseteq \Lambda$ of 0 such that Equation (2.4a) has a unique solution $\tilde{y}^*: \tilde{W} \times \Lambda_0 \longrightarrow \tilde{Y}$ which is continuous together with its first derivaties, $\tilde{y}^*(0,0), \partial\tilde{y}^*(0,0)/\partial\tilde{w} = 0$. Therefore, the solutions $x = \tilde{w} + \tilde{y}$ of Equation (2.1) which lie in a sufficiently small neighborhood of 0 for $\lambda \in \Lambda_0$ must be given by $x = \tilde{w} + \tilde{y}^*(\tilde{w},\lambda)$, where $\tilde{w}$ satisfies the equation

$$(I-\tilde{E})[B\tilde{w} - N(\tilde{w} + \tilde{y}^*(\tilde{w},\lambda))] = 0. \tag{2.5}$$

A particularly important special case of this procedure is when there is an integer p such that

$$\mathfrak{N}(B^p) = X_{\tilde{U}},\ \mathcal{R}(B^p) = Z_{\tilde{E}}. \tag{2.6}$$

If $X = Z$, a complex number μ is said to be a normal eigenvalue of a linear operator A on X if there is an integer p such that $\mathfrak{N}(A - \mu I)^p$ is finite dimensional and $X = \mathfrak{N}(A - \mu I)^p \oplus \mathcal{R}(A - \mu I)^p$. It is a fundamental result that every eigenvalue $\mu \neq 0$ of a compact linear operator is normal. Some eigenvalues may be normal even though the operator is not compact. If $X = Z$, Relation (2.6) is satisfied if 0 is a normal eigenvalue of B.

3 dim $\mathfrak{N}(B) = 1 = \operatorname{codim} \mathcal{R}(B)$, dim $\Lambda = 1$

In this section, we suppose dim $\mathfrak{N}(B) = 1 = \operatorname{codim} \mathcal{R}(B)$ and discuss bifurcation for the case in which λ is a real scalar. The next section will treat the case when λ is a vector. Suppose w_0 is a unit vector in $\mathfrak{N}(B)$, z_0 is a unit vector in $(I-E)Z$. From Lemma 2.1, we know that the bifurcation equations are one dimensional. If we define $f(u,\lambda)$ by the relation

$$\begin{aligned} z_0 f(u,\lambda) &= (I-E)N(w_0 u + y^*(w_0 u,\lambda),\lambda), \\ N(x,\lambda) &= Bx - M(x,\lambda), \end{aligned} \tag{3.1}$$

then the bifurcation equations are equivalent to the scalar equation

$$f(u,\lambda) = 0 \tag{3.2}$$

for the real scalar u and the parameter λ.

We discuss the solution set for (3.2) by imposing conditions which are computable from the known function $N(w_0 u,\lambda)$. In case λ is a scalar parameter and $f(0,\lambda) = 0$ for all λ, one obtains a very general result first stated by Crandall and Rabinowitz [19]. The proof below is different from Crandall and Rabinowitz. For another proof using the Morse Lemma, see Nirenberg [66]. See also Hale [31].

<u>Theorem 3.1</u> If $\lambda \in \mathbb{R}$, dim $\mathfrak{N}(B) = 1 = \operatorname{codim} \mathcal{R}(B)$ and N has continuous first derivatives, $\partial^2 N(x,\lambda)/\partial x \partial\lambda$ is continuous, $N(0,\lambda) = 0$ for all λ and

$$(I-E)\frac{\partial^2 N(w_0 u,\lambda)}{\partial\lambda\partial u}\Bigg|_{(0,0)} \neq 0, \tag{3.3}$$

then there is a neighborhood $V \subset X \times \mathbb{R}$ of $(0,0)$ such that the solutions $(x,\lambda) \in V$ of

$$Bx = N(x,\lambda) \tag{3.4}$$

are given by

$$x = w_0 u + y^*(w_0 u, \lambda^*(u)) \tag{3.5}$$

where $\lambda^*(u)$ is a continuously differentiable function of u with $\lambda^*(0) = 0$ and y^* is the function given in Lemma 2.1.

<u>Proof</u>. The hypotheses $N(0,\lambda) = 0$ for all λ implies $f(0,\lambda) = 0$ for all λ and the Taylor series for $f(u,\lambda)$ begins with second order terms. The condition (3.3) is equivalent to the condition $\partial^2 f(0,0)/\partial\lambda\partial u \neq 0$. Since $f(0,\lambda) = 0$ for all λ, the function $g(u,\lambda) = f(u,\lambda)/u$ is a smooth function of $(u,\lambda), g(0,0) = 0$, $\partial g(0,0)/\partial\lambda \neq 0$. The implicit function theorem and Lemma 2.1 imply the conclusion of the theorem.

Under the hypotheses of Theorem 3.1, the point $\lambda = 0$ is a bifurcation point and there is a smooth branch of solutions emanating from the line $\{(0,\lambda), \lambda \in \mathbb{R}\} \subseteq X \times \mathbb{R}$. A corollary of Theorem 3.1 is the classical result on bifurcation from simple eigenvalues of special families of functions.

<u>Corollary 3.1</u> Suppose $D: X \longrightarrow X$ has a continuous first derivative, $D(0) = 0$, $\lambda \in \mathbb{R}$. If μ_0^{-1} is a simple eigenvalue of $D'(0)$, then μ_0 is a bifurcation point for the family of functions $x - \mu Dx$. More specifically, the conclusions of Theorem 3.1 are valid.

<u>Proof</u>. For this case, $B = I - \mu_0 D'(0)$, $N(x,\lambda) = (\mu_0 + \lambda)Dx - \mu_0 D'(0)x$, $\partial^2 N(w_0 u,\lambda)/\partial\lambda\partial u\Big|_{(0,0)} = D'(0)w_0 = \mu^{-1} w_0 \neq 0$. Since μ^{-1} is a simple eigenvalue of $D'(0)$, we may take $I - E = U$ and relation (3.3) is satisfied. This proves

the corollary.

Our next objective is to determine the meaning of the condition (3.3) in the case where 0 is a normal eigenvalue of B. With $M(x,\lambda)$,B and $N(x,\lambda) = Bx - M(x,\lambda)$ as before, let

$$B_\lambda \overset{\text{def}}{=} \frac{\partial M(0,\lambda)}{\partial \lambda} \tag{3.5}$$

and suppose $\lambda = 0$ is a normal eigenvalue of B_0,

$$X = \mathfrak{N}(B_0^k) \oplus \mathfrak{R}(B_0^k)$$

for some integer k, $\dim \mathfrak{N}(B_0) = 1 = \operatorname{codim} \mathfrak{R}(B_0)$. This decomposition defines a projection P_0 whose range is $\mathfrak{N}(B_0^k)$. If $\dim \mathfrak{N}(B_0^k) = d$, then the integral formula for projections implies there is an $\varepsilon > 0$ such that for $|\lambda| < \varepsilon$, there is a continuously differentiable projection P_λ of dimension d such that $B_\lambda P_\lambda X \subseteq P_\lambda X$, $B_\lambda(I-P_\lambda)X \subseteq (I-P_\lambda)X$. Furthermore, if $\Phi_\lambda = (\phi_\lambda^1, \ldots, \phi_\lambda^d)$ is a continuously differentiable basis for $P_\lambda X$, then $B_\lambda \Phi_\lambda = \Phi_\lambda C(\lambda)$, where $C(\lambda)$ is a continuously differentiable $d \times d$ matrix function of λ. Furthermore, C(0) has only the eigenvalue 0 and $\dim \mathfrak{N}(C(0)) = 1$. Therefore, without loss in generality, we may assume

$$C(0) = \begin{bmatrix} 0 & 1 & 0 & \cdot & \cdot & \cdot & 0 \\ 0 & 0 & 1 & \cdot & \cdot & \cdot & 0 \\ \cdot & \cdot & \cdot & & & & \\ \cdot & \cdot & \cdot & & & & \\ \cdot & \cdot & \cdot & & & & \\ 0 & 0 & 0 & \cdot & \cdot & \cdot & 1 \\ 0 & 0 & 0 & \cdot & \cdot & \cdot & 0 \end{bmatrix}. \tag{3.6}$$

<u>Lemma 3.1</u> Suppose $C(\lambda)$ is defined as above. If $c_{d_1}(\lambda)$ denotes the element in the lower left hand corner of $C(\lambda)$, then

$$\frac{\partial c_{d_1}(0)}{\partial \lambda} = (-1)^d \frac{\partial}{\partial \lambda} \det C(0). \tag{3.7}$$

Proof. Expanding $\det C(\lambda)$ by cofactors along the bottom row, it is easy to see that $(-1)^d \det C(\lambda) = c_{d_1}(\lambda) - \lambda^2\gamma(\lambda)$ where $\gamma(\lambda)$ is a smooth function in $\lambda, |\lambda| < \varepsilon$. This proves the result.

In the coordinate system above, we may assume ϕ_0^1 is a basis for $\mathfrak{N}(B_0)$ and ϕ_0^d is a basis for $\mathcal{R}(B_0)$. Also,

$$(I-E)\frac{\partial^2 N(\phi_0^1 u,\lambda)}{\partial\lambda\partial u}\Bigg|_{(0,0)} = (I-E)\frac{\partial B_\lambda}{\partial\lambda}\phi_0^1\Bigg|_{(0,0)} = \frac{\partial c_{d_1}(0)}{\partial\lambda}\phi_0^d. \tag{3.8}$$

From Equation (3.8) and Theorem 3.1, we have the following result.

Corollary 3.2 If $\dim \mathfrak{N}(B) = 1 = \operatorname{codim} \mathcal{R}(B)$, if 0 is a normal eigenvalue of B and if the matrix $C(\lambda)$ is defined as above, then Relation (3.3) is satisfied if and only if

$$\frac{\partial}{\partial\lambda} \det C(\lambda)\Bigg|_{\lambda=0} \neq 0. \tag{3.9}$$

Thus, the conclusions of Theorem 3.1 are valid if Relation (3.9) is satisfied.

Corollary 3.2 includes the results of Hopf [36] and Kopell and Howard [50].

Theorem 3.1 and the resulting Corollaries 3.1, 3.2 require information only about the linear terms in x and the linear terms in λ in $M(x,\lambda)$. Without other restrictions on the nonlinearities, these results are in some sense best possible. In fact, if 0 is not a simple eigenvalue and the condition (3.9) of Corollary 3.2 is not satisfied, then there is a nonlinear perturbation for which no bifurcation occurs. We only give an example to illustrate the ideas (see Kopell and Howard [50] for the general case). Let $x = (x_1, x_2) \in \mathbb{R}^2$,

$$Dx = \begin{bmatrix} 1 & 1 \\ 0 & 1 \end{bmatrix} x + \begin{bmatrix} 0 \\ \delta x_1^2 + x_1^3 \end{bmatrix}$$

and consider the equation $M(x,\lambda) \overset{\text{def}}{=} x - (1-\lambda)Dx = 0$. For $\delta = 0$, these equations are equivalent to $x_2 = -\lambda x_1$, $[\lambda^2 + (1-\lambda)x_1^2]x_1 = 0$. If $|\lambda|$ is sufficiently small these equations have only the solution $x_1 = 0 = x_2$. Furthermore, $\partial M(x,0)/\partial\lambda = I - D'(0)$ has a one dimensional null space and has zero as a double eigenvalue. It is easy to see that $\det(I-(1-\lambda)D'(0)) = \lambda^2$ and the condition (3.9) is not satisfied, as must be the case from Lemma 3.2.

If $\delta \neq 0$, the equations in this example are equivalent to $x_2 = -\lambda x_1$, $[\lambda^2 + \delta(1-\lambda)x_1 + (1-\lambda)x_1^2]x_1 = 0$. This equation for λ small, $\lambda \neq 0$, has a nonzero solution and there is a bifurcation at $\lambda = 0$. This suggests a general result if one assumes something about the quadratic terms. We state the following result without proof. It is actually a consequence of Theorem 5.1 below.

<u>Theorem 3.2</u> If $M(x,\lambda) = I - (\mu_0 + \lambda)D(x)$ where D has a continuous second derivative, $\lambda \in \mathbb{R}$, μ_0^{-1} is a normal eigenvalue of $D'(0)$ with $\dim \mathfrak{N}(I - \mu_0 D'(0)) = 1 = \operatorname{codim} \mathcal{R}(I - \mu_0 D'(0))$ and

$$\left.\frac{\partial^2 (I-E)D(w_0 u)}{\partial u^2}\right|_{u=0} \neq 0,$$

then $\lambda = 0$ is a bifurcation point.

In this section we have considered only some very special problems in bifurcation when the family of mappings contains only one parameter.

Entire books are devoted to this subject and the following list of references will allow the reader to begin his study in this interesting area. Berger [9], Dancer [20-22], Fucik, Necas, Soucek and Soucek [24], Keller and Antman [43], Pimbley [68], Rabinowitz [69], Sather [72], Vainberg and Trenogin [88] and Westreich [91].

4 NORMAL EIGENVALUES OF ODD DIMENSIONS

In this section, we consider the special family of equations

$$x-(\mu_0+\lambda)D(x)=0, \tag{4.1}$$

where $\lambda\in\mathbb{R}$, $D:X\longrightarrow X$ has continuous first derivatives and $D(0)=0$. Also, we suppose μ_0^{-1} is a normal eigenvalue of $D'(0)$ of odd dimension d, and

$$X=\mathfrak{N}(I-\mu_0 D'(0))^k\oplus\mathcal{R}(I-\mu_0 D'(0))^k.$$

<u>Theorem 4.1</u>. Under the above hypotheses, $\lambda=0$ is a bifurcation point.

<u>Proof</u>. We apply Lemma 2.2. Since μ_0^{-1} is a normal eigenvalue, we can take $\tilde{U}=I-\tilde{E}$ and $X_{\tilde{U}}=\mathfrak{N}(I-\mu_0 D'(0))^k$. If $\Phi=(\phi_1\ldots,\phi_d)$ is a basis for $X_{\tilde{U}}$, then $X_{\tilde{U}}$ is invariant under $D'(0)$, there is a $d\times d$ matrix H whose only eigenvalue is μ_0^{-1} such that $D'(0)\ \Phi=\Phi H$. If we define $F(u,\lambda)$ uniquely by the relation

$$\Phi F(u,\lambda)=\tilde{U}[\,\Phi u-(\mu_0+\lambda)D(\Phi u+\tilde{y}^*(\Phi u,\lambda))],$$

where $u\in\mathbb{R}^d$ and $\tilde{y}^*$ is defined in Lemma 2.2, then Lemma 2.2 implies the bifurcation equations (2.5) are equivalent to

$$F(u,\lambda)=0. \tag{4.2}$$

It is a direct computation to verify that $F(0,\lambda)=0$ for all λ and that

$$\frac{\partial F(0,\lambda)}{\partial u}=(1-\lambda)I-\mu_0 H.$$

Therefore, $\lambda = 0$ is an eigenvalue of $I-\mu_0 H$ of odd multiplicity d. An elementary argument from degree theory completes the proof (see Nirenberg [66] or Krasnoselskii [54]). We do not repeat the proof here.

Theorem 4.1 for B'(0) compact is due to Krasnoselskii [54]. The above statement as well as other results can be found in MacBain [58], Ize [38].

5 $\dim \mathfrak{N}(B) = 1 = \operatorname{codim} \mathcal{R}(B)$, $\dim \Lambda > 1$

In this section, we suppose the linear operator B satisfies the same hypotheses as in Section 3 but allow the parameter λ to be a real vector of dimension greater than one. We follow the ideas of Chow, Hale and Mallet-Paret [16] and give a complete description of the bifurcations with some additional hypotheses on the nonlinearities. The method of Liapunov-Schmidt implies that we need only consider the scalar equation (3.2). In Section 1, we have discussed two particular scalar equations completely. For these examples, the bifurcation function satisfies either

$$f(0,0) = 0, \ \partial f(0,0)/\partial u = 0, \quad \partial^2 f(0,0)/\partial u^2 \neq 0, \tag{5.1}$$

or

$$f(0,0) = 0, \ \partial f(0,0)/\partial u = 0, \ \partial^2 f(0,0)/\partial u^2 = 0, \quad \partial^3 f(0,0)/\partial u^3 \neq 0. \tag{5.2}$$

If $f(u,\lambda)$ is the bifurcation function for the equation

$$M(x,\lambda) = 0, \tag{5.3}$$

then Relation (3.1) implies Relations (5.1), (5.2) are equivalent to the relations

$$N(0,0) = 0, \ \partial N(0,0)/\partial x = 0, \quad (I-E)\partial^2 N(w_0 u,0)/\partial u^2 \Big|_{u=0} \neq 0, \tag{5.4}$$

$$N(0,0) = 0, \ \partial N(0,0)/\partial x = 0, \ (I-E)\partial^2 N(w_0 u,0)/\partial u^2 = 0,$$
$$(I-E)\partial^3 N(w_0 u,0)/\partial u^3 \neq 0, \tag{5.5}$$

where $N(x,\lambda) = Bx - M(x,\lambda)$. For these two cases, it is very elementary to determine the bifurcation surfaces.

If condition (5.1) is satisfied, then in a sufficiently small neighborhood of $(u,\lambda) = (0,0)$, there is a unique $u^*(\lambda)$, $u^*(0) = 0$, such that $\partial f(u,\lambda)/\partial u = 0$ at $u = u^*(\lambda)$; that is, $f(u,\lambda)$ has a unique minimum or maximum at $(u^*(\lambda),\lambda)$. If

$$\alpha(\lambda) = f(u^*(\lambda),\lambda), \tag{5.6}$$

then, using Relation (3.1), we have the following result. (In the statement of the theorem, we assume $\alpha(\lambda)$ is a minimum. If it is a maximum, replace α by $-\alpha$).

<u>Theorem 5.1</u> If $\dim \mathfrak{N}(B) = 1 = \operatorname{codim} \mathcal{R}(B)$ and condition (5.4) is satisfied, then there is a neighborhood U of $(x,\lambda) = (0,0)$ and a continuously differentiable function $\alpha(\lambda)$, $\alpha(0) = 0$ such that the following conclusions are satisfied in U:

(i) Equation (5.3) has no solution if $\alpha(\lambda) > 0$,

(ii) Equation (5.3) has one solution if $\alpha(\lambda) = 0$,

(iii) Equation (5.3) has two distinct solutions if $\alpha(\lambda) < 0$.

For a result similar to Theorem 5.1 with applications to elliptic equations, see Ambrosetti and Prodi [2]. A more general result is contained in Gromoll and Meyer [28].

If Relation (5.2) is satisfied, then $\partial f(u,\lambda)/\partial u$ has a unique smooth minimum or maximum $\gamma_0(\lambda)$ in a sufficiently small neighborhood of $(u,\lambda) = (0,0)$. If this minimum is negative, then $f(u,\lambda)$ has a unique local maximum $\gamma_1(\lambda)$ and a unique local minimum $\gamma_2(\lambda)$ near $(u,\lambda) = (0,0)$. If $\gamma(\lambda) = \gamma_1(\lambda)\gamma_2(\lambda)$, then we have the following result. (In the statement of the Theorem, we have assumed $\gamma_0(\lambda)$ corresponds to a minimum. For a maximum, replace γ_0 by $-\gamma_0$.)

Theorem 5.2 If $\dim \mathfrak{N}(B) = 1 = \operatorname{codim} \mathcal{R}(B)$ and condition (5.5) is satisfied, then there is a neighborhood U of $(x,\lambda) = (0,0)$ and two continuously differentiable functions $\gamma_0(\lambda)$, $\gamma(\lambda)$ vanishing at $\lambda = 0$ such that the following conclusions are satisfied in U:

(i) If $\gamma_0(\lambda) \geqslant 0$ there is a unique solution of Equation (5.3),

(ii) If $\gamma_0(\lambda) < 0$ then $\gamma(\lambda)$ is defined and

(a) $\gamma(\lambda) > 0$ implies one simple solution of Equation (5.3),

(b) $\gamma(\lambda) = 0$ implies one simple and one double solution of Equation (5.3),

(c) $\gamma(\lambda) < 0$ implies three simple solutions of Equation (5.3).

Theorems 5.1 and 5.2 show that the phenomena discovered in the two examples of the introduction are typical in the case where $\dim \mathfrak{N}(B) = 1 = \operatorname{codim} \mathcal{R}(B)$. The bifurcation surface under condition (5.4) is $\alpha(\lambda) = 0$ and under condition (5.5) is $\gamma(\lambda) = 0$. These functions may be computed to any accuracy desired from the relation

$$z_0 f(u,\lambda) = (I-E)N(w_0 u + y^*(w_0 u,\lambda),\lambda),$$

$$N(x,\lambda) = Bx - M(x,\lambda),$$

and, therefore, one can obtain the approximate bifurcation curves.

Note that the parameter λ can be infinite dimensional and therefore these theorems can be applied to obtain the complete bifurcation picture for all maps S in a neighborhood of the given map $M(\cdot,0)$ with the topologies on the maps being the C^r-topology. The parameter λ can be chosen to be $\lambda = S - M(\cdot,0)$. Although this direction is interesting to pursue, we do not discuss it any further.

In case the family of mappings $M(\cdot,\lambda)$ has certain special forms, the approximate values of the functions in Theorems 5.1 and 5.2 are easy to obtain. For example, if

$$M(\cdot,\lambda) = I-(a+\lambda_1)F - \lambda_2 G, \tag{5.7}$$

where F,G have continuous third derivatives,

$$\begin{aligned}
&a \neq 0,\\
&F(0) = 0,\\
&\dim \mathfrak{N}(I-a F'(0)) = 1 = \operatorname{codim} \mathcal{R}(I-a F'(0)),\\
&[w_0] = \mathfrak{N}(I-a F'(0)),\\
&[z_0] = \operatorname{coker}[I-a F'(0)],
\end{aligned} \tag{5.8}$$

where [ ] denotes span, then $B = I-a F'(0)$, $N(x,\lambda) = Bx-M(x,\lambda) = a[F(x)-F'(0)x] + \lambda_1 F(x) + \lambda_2 G(x)$.

The conditions (5.4), (5.5) for this special family are

$$\begin{aligned}
&F(0) = 0,\ (I-E)F'(0)w_0 = 0,\\
&z_0\varepsilon \stackrel{\text{def}}{=} (I-E)F''(0)(w_0,w_0) \neq 0,
\end{aligned} \tag{5.9}$$

$$\begin{aligned}
&F(0) = 0,\ (I-E)F'(0) = 0,\ (I-E)F''(0)(w_0,w_0) = 0,\\
&z_0\delta \stackrel{\text{def}}{=} (I-E)F'''(0)(w_0,w_0,w_0) \neq 0.
\end{aligned} \tag{5.10}$$

If condition (5.9) is satisfied, then $\alpha(\lambda)$ in Theorem 5.1 is given by

$$\alpha(\lambda) = \nu\lambda_2 - \frac{(\mu\lambda_1)^2}{4\varepsilon} + \text{h.o.t.},\quad z_0\mu = (I-E)F'(0)w_0,\ z_0\nu = (I-E)G(0) \tag{5.11}$$

where h.o.t. denotes higher order terms near $(\lambda_1,\lambda_2) = (0,0)$. The bifurcation surface is given approximately by (see Figure 5)

$$\nu\lambda_2 = (\mu\lambda_1)^2/4\varepsilon . \tag{5.12}$$

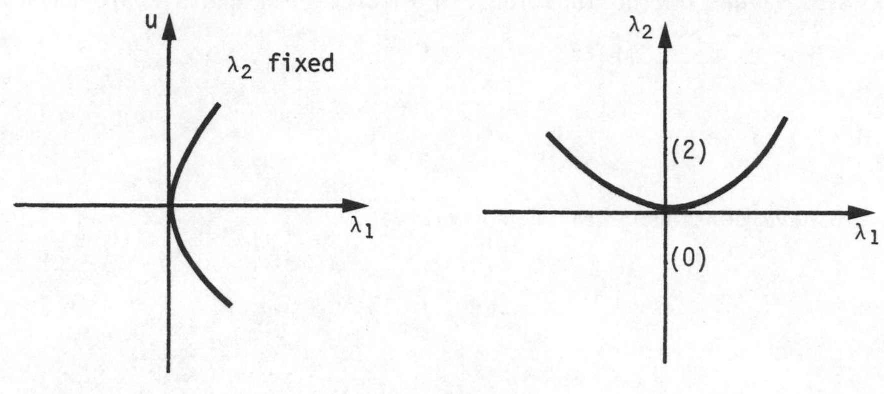

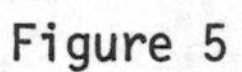

Figure 5

If condition (5.10) is satisfied, then $\gamma(\lambda)$ in Theorem 5.2 is given by

$$\gamma(\lambda) = -\frac{4}{27}\left(\frac{\mu\lambda_1}{\delta}\right)^3 + \left(\frac{\nu\lambda_2}{\delta}\right)^2 + \text{h.o.t.} \tag{5.13}$$

The bifurcation surface is then given approximately by the cusp (see Figure 6),

$$\frac{4}{27}\,\frac{(\mu\lambda_1)^3}{\delta} = (\nu\lambda_2)^2. \tag{5.14}$$

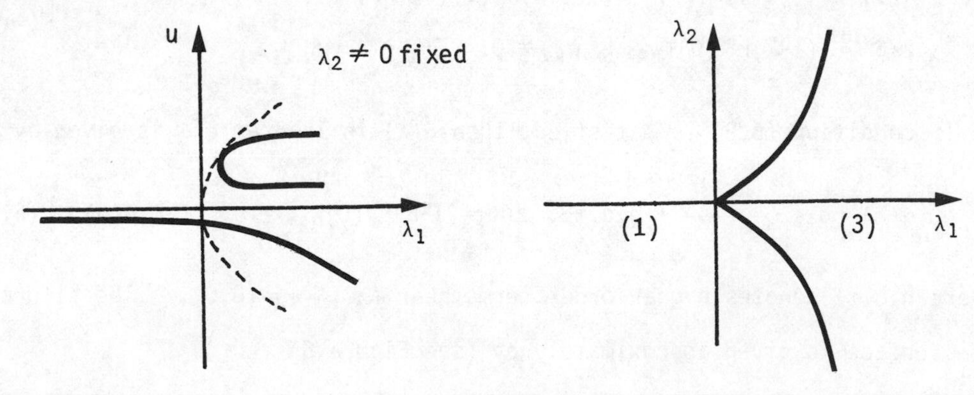

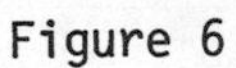

Figure 6

The dotted curve (parabola) in Figure 6 represents the locus of the double solution as one traverses the cusp. It is approximately given by $3\delta u^2 = \mu\lambda_1$.

6 SOME APPLICATIONS

In this section, we give some applications of the previous results from Chow, Hale and Mallet-Paret [16] to the buckling problem for plates and to the non-uniqueness of solutions of some elliptic boundary value problems. Since the buckling problem will be used as an illustration for many of the later results, we introduce the equations in a more general form than needed here. These equations also serve as a model for interesting families of abstract mappings $M(x,\lambda)$ to be discussed. Our discussion centers on rectangular plates but other shapes could also be considered.

Suppose the von Kármán equations describe the buckling of a plate or shell under external disturbances. Suppose the plate is described by the rectangular domain

$$\Omega = (0,\ell) \times [0,1] \subseteq \mathbb{R}^2$$

in the (x,y)-plane. In the absence of external forces, suppose the plate is not perfectly flat so that it has a small imperfection which represents a displacement $\alpha w_0(x,y)$, where α is a small parameter and $w_0: \Omega \to \mathbb{R}$ is a known function. Let λ be a lateral force applied to the edges of the plate at $x = 0$, $x = \ell$, F be the Airy stress function and let $\nu h(x,y)$ be a vertical load on the plate, where ν is a small parameter and h is a known function. If w and f are the additional deflection and stress caused by the normal loading, the vertical loading and the imperfection, the von Kármán equations are

$$\begin{aligned} &\Delta^2 f = -\tfrac{1}{2}[w,w] - \alpha[w,w_0], \\ &\Delta^2 w = [w + \alpha w_0,\ f + \lambda F], \quad \text{in } \Omega \end{aligned} \tag{6.1}$$

where $[u,v] = u_{xx}v_{yy} + u_{yy}v_{xx} - 2u_{xy}v_{xy}$ and Δ is the Laplace operator. For a simply supported plate, the boundary conditions are

$$f = 0, \ \Delta f = 0, \ w = 0, \ \Delta w = 0, \quad \text{on } \partial\Omega. \tag{6.2}$$

The final shape of the plate is given by $\alpha w_0(x,y) + w(x,y)$. For further explanation of the equations, see Berger [8], Berger and Fife [10], Knightly and Sather [45,46], Knightly [47], Sather [73].

As in Knightly and Sather [45], we may write (6.1), (6.2) as an operator equation in the Hilbert space

$$X = \{u \in H^2(\Omega) : u = 0 \text{ on } \partial\Omega\},$$
$$\langle u,v \rangle = \int_\Omega (\Delta u)(\Delta v).$$

If Δ^{-1} is the inverse of the Laplacian with zero initial data and, if for $u,v \in X$, we define

$$Lu = \Delta^{-2}[u,F] = -\Delta^{-2}u_{xx},$$
$$\Lambda u = \Delta^{-2}[u,w_0],$$
$$B(u,v) = \Delta^{-2}[u,v],$$
$$Q(u) = B(u,\Lambda u) + \tfrac{1}{2}\Lambda B(u,u),$$
$$C(u) = \tfrac{1}{2}B(u,B(u,u)),$$
$$p = -\Delta^{-2}w_{0xx},$$
$$g = -\Delta^{-2}h,$$

then Equations (6.1), (6.2) are equivalent to

$$C(w) + \alpha Q(w) + (I - \lambda L + \alpha^2\Lambda^2)w - \alpha\lambda p + \nu g = 0. \tag{6.3}$$

One is interested in studying the behavior of the solutions of this equation for (λ,α,ν) near $(\lambda_0,0,0)$ where λ_0^{-1} is the first eigenvalue of L. If ℓ, the length of the edge of the plate in the x-direction, is not $\sqrt{2}$, then λ_0^{-1} is a simple eigenvalue of L. Thus, the theory of the previous sections is applicable. If $\ell=\sqrt{2}$, λ_0^{-1} is a double eigenvalue. This case will be discussed later. To completely understand the bifurcations in this latter case one must also allow ℓ to be an independent parameter varying near $\sqrt{2}$.

In the remainder of the discussion for this example suppose $\ell\neq\sqrt{2}$, so that λ_0^{-1} is a simple eigenvalue of L, and also suppose $\alpha=0$. Then $\dim \mathfrak{N}(I-\lambda_0 L)=1=\operatorname{codim} \mathcal{R}(I-\lambda_0 L)$ and one may choose the projections U,E onto $\mathfrak{N}(I-\lambda_0 L)$ and $\mathcal{R}(I-\lambda_0 L)$, respectively, so that $U=I-E$. The resulting bifurcation equations have the form

$$f(u,\lambda_1,\lambda_2)=c_0u^3+c_1\lambda_1 u+c_2\lambda_2+\text{h.o.t.}=0,$$
$$\text{h.o.t.}=O(u^4+|\lambda_1 u^2|+|\lambda_1^2 u|+\lambda_2^2+|\lambda_1\lambda_2|+|\lambda_2 u|),$$

as $u,\lambda_1=(\lambda-\lambda_0)$, $\lambda_2=\nu$ approach zero, where c_0,c_1,c_2 are constants which are computed as described in Section 5. In fact, if ζ is a basis vector for $\mathfrak{N}(I-\lambda_0 L)$, then

$$\begin{aligned}&c_0\zeta=UC(\zeta) \text{ and } c_0>0,\ c_1=-\lambda_0^{-1},\\ &c_2\zeta=-Uq.\end{aligned} \tag{6.4}$$

The results of Section 5 are immediately applicable and one obtains the following result (see also Keener and Keller [40]).

Theorem 6.1 Let λ_0 be a simple eigenvalue of the linear operator L in Equation (6.3) and $c_2\neq 0$ in Relation (6.4). Then there is a bifurcation of Equation (6.3) from zero. Moreover, there exist two functions $\gamma_0(\lambda,\nu)$ and $\gamma(\lambda,\nu)$ as in Theorem 5.2 such that for $|\lambda-\lambda_0|,|\nu|$ and $|u|$ small the following hold:

(i) the functions γ_0 and γ determine a unique curve $\lambda(\nu)$, denoted by Γ, in $\lambda\nu$-plane (see Figure 6). The curve is smooth when $\nu \neq 0$ but forms a cusp tangent to the λ axis at the point $(\lambda_0, 0)$. In fact, $\lambda(\nu)-\lambda_0 \sim K\nu^{2/3}$ where $K = 3\lambda_0 \left(\frac{c_0 c_2^2}{4}\right)^{1/3}$,

(ii) $(\lambda,\nu) \notin \Gamma$ for any $\lambda < \lambda_0$,

(iii) $(\lambda,\nu) \in \Gamma, \lambda \neq \lambda_0$ implies there is exactly one nontrivial solution of (6.3),

(iv) Γ divides a neighborhood of $(\lambda_0, 0)$ into two components (open and connected) Γ_1 and Γ_2 such that

(a) $(\lambda,\nu) \in \Gamma_1 \Rightarrow$ there is a unique solution of (6.3),

(b) $(\lambda,\nu) \in \Gamma_2 \Rightarrow$ there are exactly three solutions of (6.3), all distinct.

As mentioned before, one may compute the approximate quantities γ_0, γ_1 in terms of the λ_0-eigenfunction of (6.3) and obtain the asymptotic form of $\lambda(\nu)$ above, thus determining the nature of the bifurcation. The specific type of information that is obtained may also be depicted in a manner similar to that depicted in Figure 6.

For an interesting discussion of the plate problem with no normal loading and large lateral force, see Ambrosetti [2]. The same problem as above is discussed using catastrophe theory by Chillingworth [14].

The hypothesis $c_2 \neq 0$ in Theorem 6.1 is generic in the sense that most normal loadings will satisfy this property. However, if the external load satisfies some symmetry properties this constant is equal to zero and the theorem does not apply. Vanderbauwhede [89] has discussed the bifurcation problem in this case. He has also described the manner in which the transition can take place between the generic and the symmetric nongeneric case.

The results of Sattinger [74] could possibly be used effectively in other problems with symmetry.

As another example, let us consider the problem of Keener [39] (see also D. S. Cohen [18]) on the branching phenomena for a nonlinear elliptic boundary value problem of the form

$$\begin{aligned} Lu + N(\lambda,\mu,u) &= 0, \quad x \in G, \\ Bu &= 0, \quad x \in \partial G \end{aligned} \tag{6.5}$$

where

G is a bounded open set in $\mathbb{R}^n$

∂G is the smooth boundary of G,

L is a uniformly elliptic second order operator (with or without constant coefficients),

N is the nonlinearity, smooth in its arguments,

B is a boundary operator,

λ,μ are parameters.

We assume that

$$N(\lambda,0,0)=0 \quad \text{for all real } \lambda \text{ (the unforced case)},$$
$$N(\lambda,\mu,0)\neq 0 \quad \text{if } \mu\neq 0 \text{ (the forced case)}. \tag{6.8}$$

In many applications, the nonlinearity N satisfies

$$\frac{\partial^m}{\partial u^m} N(\lambda_0,0,0)\equiv 0, \quad 2\leqslant m<k, \quad \frac{\partial^k}{\partial u^k} N(\lambda_0,0,0)\neq 0 \tag{6.9}$$

where λ_0 is such that the following boundary value problem has one dimensional generalized kernel and cokernel:

$$Lu+\frac{\partial}{\partial u} N(\lambda_0,0,0)u=0,$$
$$Bu=0. \tag{6.10}$$

By interpreting (6.5) as an equation in an appropriate Banach space, one can apply the above theory to obtain the bifurcation equation

$$f(u,\lambda,\mu)=0 \tag{6.11}$$

where $u\in\mathbb{R}$ and $f:\mathbb{R}^3\to\mathbb{R}$ is smooth.

The case $k=2$ follows from Theorem 5.1. The bifurcation curve is described approximately by the Relation (5.11) and is depicted in Figure 6.

For $k=3$, the bifurcation phenomena of (6.3) are similar to those in von Kármán's equation provided the generic conditions are satisfied.

Suppose now $k>3$. This is a nongeneric case in the sense that the family of equations (6.11) consists of two parameters and the degeneracy in $f(u,0)$ is of order >3. However, it is possible to compute the catastrophe set under some hypotheses on $f(u,\lambda,\mu)$. Without loss in generality, take $\lambda_0=0$.

Suppose that

$$f_\mu(0,0,0) \neq 0, \tag{6.12}$$

$$f(0,\lambda,0) \equiv 0, \quad f_{\lambda u}(0,0,0) \neq 0, \tag{6.13}$$

$$f(u,0,0) = \alpha u^k + O(|u|^{k+1}), \quad \alpha \neq 0. \tag{6.14}$$

Under these hypotheses, the Jacobian

$$\begin{pmatrix} f_\mu(0,0,0) & f_{\mu u}(0,0,0) \\ f_\lambda(0,0,0) & f_{\lambda u}(0,0,0) \end{pmatrix} = \begin{pmatrix} \neq 0 & * \\ 0 & \neq 0 \end{pmatrix}$$

is non-singular. By the implicit function theorem there are unique solutions $\lambda = \lambda(u)$, $\mu = \mu(u)$ to

$$\lambda = C_1 u^{k-1} + O(|u|^k), \quad C_1 < 0 \text{ constant}$$

$$\mu = C_2 u^k + O(|u|^{k+1}), \quad C_2 > 0 \text{ constant.}$$

This curve in the (λ,μ)-plane represents the occurrence of double zeros, and thus parameter values at which the number of solutions of $f = 0$ changes. Except at the origin, none of these points represents triple (or higher order) zeros since

$$f_{uu}(u,\lambda(u),\mu(u)) \sim Cu^{k-2} \neq 0.$$

If we return to the original equation (6.3) and assume that

$$\pi \frac{\partial}{\partial \mu} N(\lambda_0,0,0) \neq 0,$$

$$\pi \frac{\partial}{\partial \lambda} \frac{\partial}{\partial u} N(\lambda_0,0,0) w_0 \neq 0,$$

$$\pi \frac{\partial^k}{\partial u^k} N(\lambda_0,0,0) w_0^k \neq 0;$$

then (6.12), (6.13) and (6.14) are satisfied. Here w_0 is the unique (up to constant multiple) nonzero solution of (6.10) and π the projection onto the cokernel of this linear operator.

7 Normal Forms, $\dim \mathfrak{N}(B) = 1 = \operatorname{codim} \mathcal{R}(B)$

The results of Section 5 are special cases of a much more general theory; namely, catastrophe theory or the theory of normal forms or universal unfoldings of a mapping. For the case where $\dim \mathfrak{N}(B) = 1 = \operatorname{codim} \mathcal{R}(B)$, we have seen that the solution of the bifurcation problem reduces to the discussion of the manner in which the solutions u of a scalar equation $f(u,\lambda) = 0$ depend on λ. Under some hypotheses on the Taylor series for $f(u,0)$ (the second degree term does not vanish or the third degree term does not vanish), we gave a complete solution to this problem in the introduction and Section 5. If the function $f(u,\lambda)$ is analytic in u,λ, then the Weierstrass preparation theorem says this problem is equivalent to a polynomial of degree k if the first nonzero term in the Taylor series of $f(u,0)$ is of degree k. In the nonanalytic case, a similar result is valid and relies heavily upon the Malgrange preparation theorem. The following result of Thom which will not be proved is a precise statement of these remarks. For theoretical and practical implications, see Arnold [4], Golubitsky and Guillemin [26], Koiter [48,49], Kuiper [55], Sewell [77], Takens [79-82], Thom [83,84], Thom and Zeeman [85], Thompson and Hunt [89], Wasserman [90], and the collection of papers in [23].

Theorem 7.1 Suppose $f: \mathbb{R} \times \mathbb{R}^n$ is a C^∞-map and $f(u,\lambda)$ satisfies

$$f(u,0) = cu^k + O(|u|^{k+1}), \; c \neq 0, \tag{7.1}$$

$$\operatorname{rank}\,(a_{ij}) = k-1 \text{ where } \frac{\partial f}{\partial \lambda_i}(u,0) = \sum_{j=0}^{k-2} \frac{a_{ij}}{j!} u^j + O(|u|^{k-1}). \tag{7.2}$$

Then there exists a C^∞ change of coordinates $\overline{\lambda} = \eta(\lambda) \in \mathbb{R}^{k-1}$, $\overline{\xi} = \xi(u,\lambda) \in \mathbb{R}$ near $(u,\lambda) = (0,0)$ such that, in the new coordinates, f has the form

$$\overline{f}(\overline{u},\overline{\lambda}_1,\ldots,\overline{\lambda}_{k-1}) = \overline{u}^k + \sum_{i=0}^{k-2} \overline{\lambda}_{i+1}\,\overline{u}^i. \tag{7.3}$$

The function $\overline{f}(\overline{u},\overline{\lambda})$ in Equation (7.3) is called the normal form of $f(u,\lambda)$ or the universal unfolding of the function f. Condition (7.2) says that the parameters λ enter in the function f in such a way as to influence all terms in the power series of degree $\leqslant k-2$. Note that the coefficient in $\overline{u}^{k-1}$ is absent from (7.3). From our point of view, this is no restriction since one can always make a translation of variables to eliminate this term.

The function $\overline{f}(\overline{u},\overline{\lambda})$ carries all of the information necessary to solve the bifurcation problem. It contains only a finite number of parameters $\overline{\lambda}$ and the set in $\overline{\lambda}$-space across which the number of solutions $\overline{u}$ of $\overline{f}(\overline{u},\overline{\lambda}) = 0$ changes is called a bifurcation surface or catastrophe set. Once these sets are known, one can recover the corresponding bifurcation surfaces in the original parameter space λ by using the function $\overline{\lambda} = \eta(\lambda)$. Of course, η will never be known exactly but can be obtained approximately from the matrix (a_{ij}) in Relation (7.2).

To see Theorem 7.1 for some special cases, suppose $\lambda \in \mathbb{R}$ and

$$f(u,0) = cu^2 + O(|u|^3), \quad c \neq 0, \tag{7.4}$$

$$\frac{\partial f(0,0)}{\partial\lambda} \neq 0.$$

Condition (7.1) is satisfied for $k = 2$ and condition (7.2) is satisfied. Thus, Theorem 7.1 implies there is a C^∞ change of coordinates $\overline{u} = \xi(u,\lambda) \in \mathbb{R}$, $\overline{\lambda} = \eta(\lambda) \in \mathbb{R}$ such that the normal form for f is given by

$$\overline{f}(\overline{u},\overline{\lambda}) = \overline{u}^2 + \overline{\lambda}.$$

The bifurcation surface is therefore $\overline{\lambda} = 0$ with no solutions for $\overline{\lambda} > 0$ and two solutions for $\overline{\lambda} < 0$. Also, $c\overline{\lambda} = (\partial f(0,0)/\partial\lambda)\lambda + O(|\lambda|^2)$.

If $\lambda \in \mathbb{R}^2$, $f(u,\lambda)$ satisfies Relation (7.4) and is the bifurcation function of maps in Relation (5.7), then $c\overline{\lambda} = \nu\lambda_2 - (\mu\lambda_1)^2/4 + \text{h.o.t.}$ as in Equation (5.11).

If $\lambda \in \mathbb{R}^2$ and

$$f(u,0) = cu^3 + O(|u|^4), \quad c \neq 0,$$

$$\frac{\partial^2 f(0,0)}{\partial\lambda_1 \partial u} \neq 0, \quad \frac{\partial f(0,0)}{\partial\lambda_2} \neq 0,$$

then Condition (7.1) is satisfied for $k = 3$ and Condition (7.2) is satisfied. Therefore, there exists a C^∞ change of coordinates $\overline{u} = \xi(u,\lambda) \in \mathbb{R}$, $\overline{\lambda} = \eta(\lambda) \in \mathbb{R}^2$ such that the normal form for f is

$$\overline{f}(\overline{u},\overline{\lambda}) = \overline{u}^3 + \overline{\lambda}_1\overline{u} + \overline{\lambda}_2.$$

The bifurcation surface is the familiar cusp and is obtained approximately in (λ_1,λ_2)-space through the relations

$$c\overline{\lambda}_1 = \lambda_1 \partial^2 f(0,0)/\partial\lambda_1\partial u + O(\lambda_1^2),$$

$$c\overline{\lambda}_2 = \lambda_2 \partial f(0,0)/\partial\lambda_2 + O(\lambda_2^2).$$

For $f(u,0)$ nondegenerate of order k as in Relation (7.1), one needs at least k-1 parameters to describe the normal form for f. Therefore, the bifurcation surfaces are in very high dimensional spaces. In a physical problem, one may be interested in determining the nature of bifurcation when fewer than k-1 parameters are varied; say, for example, two parameters. This means that we must determine how a plane intersects the high dimensional bifurcation surface - a nontrivial problem. In Section 6, we gave an example of two parameters when $f(u,0)$ was degenerate of order k. We were able to obtain the bifurcation surfaces from elementary calculus without going through the normal form. Of course, the family of functions were special and we obtained much less information than is contained in the normal form. Most of the discussion in later sections will proceed from the point of view of elementary calculus for special families rather than use the normal form. The main reason for doing this is that the number of parameters needed to describe the normal forms for mappings from $f: \mathbb{R}^2 \times \mathbb{R}^n \to \mathbb{R}^2$ (which would correspond to bifurcation functions for $\dim \mathfrak{N}(B) = 2 = \operatorname{codim} \mathcal{R}(B)$) increase very rapidly with the degeneracy. For example, if the two vector $f(u,0)$ is two homogeneous cubics in $u = (u_1, u_2)$, one needs at least eight parameters. Furthermore, the theory of normal forms is only valid for gradient systems.

Of course, it should be obvious that the theory of normal forms will always play an important role in the applications and will be used more frequently as the theory develops. So much qualitative information is contained in these normal forms that they cannot be neglected. We give one illustration in the next section to a problem in ordinary differential equations.

8 GENERALISED HOPF BIFURCATION

In this section, we apply the results on normal forms to obtain the interesting results of Takens [79] on generalized Hopf bifurcation. Consider the second order equation in the plane, $x = (x_1, x_2)$,

$$\dot{x} = Ax + X(x,\mu), \quad A = \begin{bmatrix} 0 & 1 \\ -1 & 0 \end{bmatrix}, \tag{8.1}$$

where $\mu \in \mathbb{R}^n$ is a parameter, X is C^∞ and satisfies

$$X(0,\mu) = 0, \quad \frac{\partial X(0,0)}{\partial x} = 0, \quad \frac{\partial X(0,\mu)}{\partial \mu} = 0. \tag{8.2}$$

Introduce polar coordinates (ρ,Θ) in (8.1) and eliminate t to obtain the scalar equation

$$\frac{d\rho}{d\Theta} = R(\rho,\Theta,\mu) \tag{8.3}$$

where $R(\rho,\Theta,\mu) = R(\rho,\Theta + 2\pi,\mu)$. Obtaining 2π-periodic solutions of (8.3) is equivalent to finding periodic solutions of (8.1). One now applies the classical procedure for obtaining the bifurcation equations for the periodic solutions of (8.3) (see [30]) in the form

$$0 = f(a,\mu) = \frac{1}{2\pi}\int_0^{2\pi} R(a + \rho^*(\Theta,a,\mu),\Theta,\mu)\,d\Theta \tag{8.4}$$

where a is a scalar and $\rho^*(\Theta,a,\mu)$ is a 2π-periodic function of mean value zero which satisfies

$$\frac{d\rho}{d\Theta} = R(\rho,\Theta,\mu) - f(a,\mu).$$

The function $f(a,\mu)$ is C^{∞} and, from the manner in which it is constructed, one can show it is odd in a. Then $f(a,\mu)/a = g(a^2,\mu)$, where $g(r,\mu)$ is C^{∞}. Suppose

$$g(r,0) = r^k + O(r^{k+1}). \tag{8.5}$$

Theorem 7.1 says that $g(r,\mu)$ is equivalent to

$$g(r,\mu) = r^k + \alpha_1(\mu)r^{k-1} + \ldots + \alpha_k(\mu)$$

in a neighborhood of $r = 0, \mu = 0$. Therefore, $f(a,\mu)$ is equivalent to

$$f(a,\mu) = ag(a^2,\mu) = a^{2k+1} + \alpha_1(\mu)a^{2k-1} + \ldots + \alpha_2(\mu)a.$$

Thus, the manner in which the periodic solutions of (8.1) appear and disappear as a function of μ can be ascertained. The pictures are in Takens' paper [79]. For related work, see Chafee [13].

We remark that another proof of the classical Hopf bifurcation theorem is immediate from (8.4) and the same argument as used for simple eigenvalues in Section 3. For an extensive bibliography as well as applications of simple Hopf bifurcation, see Marsden and MacCracken [64]. For the global Hopf bifurcation, see Alexander and Yorke [1], Ize [37], Chow and Mallet-Paret [15], Nussbaum [67].

9 THE CUSP REVISITED

In this section, we use a different method for discussing the bifurcations in the scalar equation

$$f(u,\lambda) = 0, \quad f \in \mathbb{R},\ u \in \mathbb{R},\ \lambda \in \mathbb{R}^2, \tag{9.1}$$

when $\lambda = (\lambda_1, \lambda_2)$ and

$$f(u,0) = u^3 + O(|u|^4), \tag{9.2}$$

$$\frac{\partial^2 f(0,0)}{\partial\lambda_1 \partial u} \neq 0, \quad \frac{\partial f(0,0)}{\partial \lambda_2} \neq 0.$$

The method is very elementary and consists only of the systematic application of scaling techniques and the implicit function theorem (see Chow, Hale and Mallet-Paret [17]. However, the method is easily extended to certain types of bifurcation problems in higher dimensions and the ideas can be adapted to problems of an entirely different type.

Condition (9.2) implies that

$$f(u,\lambda) = u^3 + \alpha_1\lambda_1 u + \alpha_2\lambda_2 + \text{h.o.t.}, \tag{9.3}$$

where $\alpha_1 \neq 0$, $\alpha_2 \neq 0$, and h.o.t. $= O(|u|^4 + |\lambda_1 u^2| + |\lambda_1|^2 + |\lambda_1\lambda_2| + |\lambda_2|^2 + |\lambda_2 u|)$ as u, λ_1, λ_2 approach zero.

The fact that the u^3 term appears in the Taylor series expansion of $f(u,\lambda)$ permits one to obtain a priori bounds for all solutions of Equation (9.1) in a neighborhood of $(u,\lambda) = (0,0)$. In fact, one can prove

Lemma 9.1 There is a neighborhood V of $(u,\lambda) = (0,0)$ and a constant $\beta \neq 0$ such that any solution of Equation (9.1) in V must satisfy

$$|u| \leq \beta(|\lambda_1|^{1/2} + |\lambda_2|^{1/3}).$$

Proof: If this is not the case then there exist a sequence of solutions $\{(u_n,\lambda_{1n},\lambda_{2n})\}$ approaching zero such that $|\lambda_{1n}|^{1/2}/|u_n| \longrightarrow 0$, $|\lambda_{2n}|^{1/3}/|u_n| \longrightarrow 0$ as $n \longrightarrow \infty$. We divide Equation (9.1) by $|u_n|^3$ to obtain

$$0 = \frac{f(u_n,\lambda_n)}{|u_n|^3} = 1 + 0\left(|u_n| + \frac{|\lambda_{1n}|}{|u_n|^2} + \frac{|\lambda_{2n}|}{|u_n|^3}\right),$$

which gives a contradiction since the right hand side approaches one as $n \to \infty$. This proves the lemma.

Lemma 9.1 justifies certain scalings which will have the effect of reducing the two parameter problem to essentially a one parameter problem. In fact, if

$$u = \lambda_2{}^{1/3}v, \quad \lambda_1 = \lambda_2{}^{2/3}\mu, \tag{9.4}$$

then Equation (9.1) becomes equivalent to the equation

$$v^3 + \alpha_1\mu v + \alpha_2 = 0(|\lambda_2|^{1/3}). \tag{9.5}$$

If we consider Equation (9.4) for all values of $\mu \in \mathbb{R}$, $v \in \mathbb{R}$, and λ_2 small, then Lemma 9.1 implies no solutions are lost by the transformation (9.4). Thus, Equation (9.5) is actually equivalent to Equation (9.1).

To analyze Equation (9.5), we first observe that Hypothesis (9.2) implies $\alpha_2 \neq 0$. Therefore, all solutions of the equation

$$v^3 + \alpha_2 = 0$$

are simple. The implicit function theorem implies that all solutions of Equation (9.5) are simple for $|\mu| < \delta$, $|\lambda_2| < \delta$ for some $\delta > 0$. Consequently, no bifurcations can occur in this region. Returning to the (λ_1, λ_2)-space, this means that no bifurcations occur near the λ_2-axis in a region as shown in Figure 7.

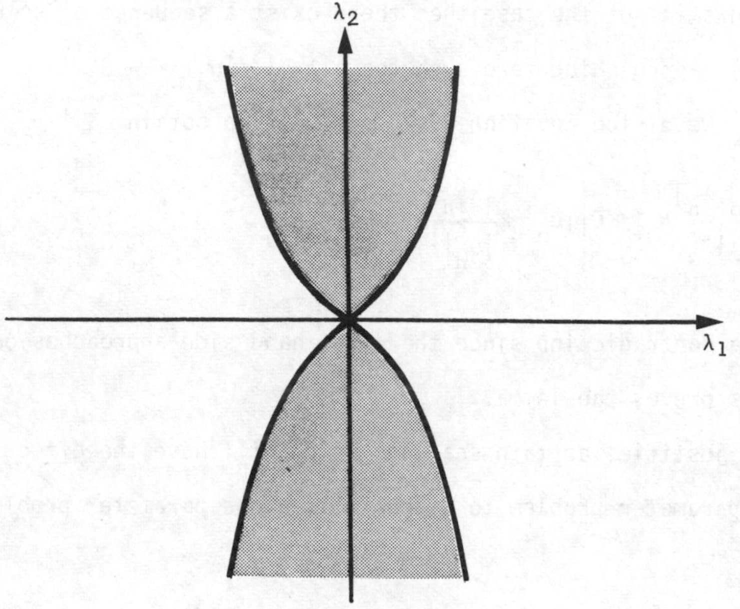

Figure 7

From this elementary observation, we need only consider Equation (9.5) for $|\mu| \geqslant \delta > 0$, $v \in \mathbb{R}$ and λ_2 small. To avoid working with the unbounded set $|\mu| \geqslant \delta$, it is more convenient to parametrize the neighborhood of $(\lambda_1, \lambda_2) = (0,0)$ by representing λ_2 as a function of λ_1 rather than λ_1 as a function of λ_2 as in Relation (9.4). If this is done, then u should also be parametrized by λ_1 and the appropriate scaling is again given by Lemma 9.1. In fact, it is justified exactly as before to let

$$u = |\lambda_1|^{\frac{1}{2}} v, \quad \lambda_2 = |\lambda_1|^{3/2} \nu, \tag{9.6}$$

in order to obtain the equivalent equation

$$v^3 \pm \alpha_1 v + \nu\alpha_2 = O(|\lambda_1|^{\frac{1}{2}}) \qquad (9.7)$$

where the + sign corresponds to $\lambda_1 > 0$ and the minus sign to $\lambda_1 < 0$. For Equation (9.7), we need only consider all solutions for λ_1 small and ν in a bounded set.

The bifurcation curves for Equation (9.7) are determined by the multiple solutions. Thus, we must consider simultaneous solutions of Equation (9.7) and the equation

$$3v^2 \pm \alpha_1 = O(|\lambda_1|^{\frac{1}{2}}). \qquad (9.8)$$

Hypothesis (9.2) implies $\alpha_1 \neq 0$. Therefore, any solutions (v,ν) of Equation (9.7), (9.8) for $\lambda_1 = 0$ must have $v \neq 0$. Furthermore, the Jacobian of the left hand sides of these equations with respect to (v,ν) is given by $-6\alpha_2 v$ which is different from zero at any solution (v_0,ν_0) of Equations (9.7), (9.8) for $\lambda_1 = 0$. The implicit function theorem implies there exist unique solutions $v^*(\lambda_1)$, $\nu^*(\lambda_1)$ for λ_1 sufficiently small which satisfy $v^*(0) = v_0$, $\nu^*(0) = \nu_0$. In the original parameter space (u,λ), this means that the cusp $\lambda_2 = \lambda_2^*(\lambda_1) = |\lambda_1|^{\frac{1}{2}}\nu^*(\lambda_1)$ is a good candidate for the bifurcation curve and the value of the solution u on this curve is given by $u^*(\lambda_1) = |\lambda_1|^{\frac{1}{2}}v^*(\lambda_1)$.

For this curve to be a bifurcation curve, one shows that the number of solutions change by two as the curve is crossed. This is a consequence of the fact that the first derivative of the left side of Equation (9.8) with respect to v is 6v and the first derivative of the left hand side of Equation (9.7) with respect to ν is α_2, which are different from zero along the curve $\nu = \nu^*(\lambda_1)$.

The previous remark also shows that Equations (9.7), (9.8) have exactly one solution and thus there is only one bifurcation curve.

By this procedure, we have made a complete analysis of the bifurcations. The procedure is more complicated than the one given in Section 1. It has the advantage of leading to generalizations to higher dimensions. This is illustrated in the next section.

10 BIFURCATION IN TWO DIMENSIONS. CUBIC NONLINEARITIES AND TWO PARAMETERS

In this section, we consider the bifurcation of solutions of the equation

$$\begin{aligned} &f(u,\lambda) = 0, \\ &f \in \mathbb{R}^2,\ u \in \mathbb{R}^2,\ \lambda \in \mathbb{R}^2, \end{aligned} \tag{10.1}$$

where for $u = (u_1, u_2)$, $\lambda = (\lambda_1, \lambda_2)$, $f(u,\lambda)$ has the form

$$f(u,\lambda) = C(u) + \lambda_1 Lu + \lambda_2 k + \text{h.o.t.}, \tag{10.2}$$

in which

$$\begin{aligned} &C: \mathbb{R}^2 \to \mathbb{R}^2 \text{ is a homogeneous cubic,} \\ &L: \text{is a } 2 \times 2 \text{ matrix,} \\ &k \in \mathbb{R}^2 \text{ is given,} \\ &\text{h.o.t.} = O(|u|^4 + |\lambda_1 u|^2 + |\lambda_1|^2 + |\lambda_1 \lambda_2| + |\lambda_2|^2 + |\lambda_2 u|) \end{aligned} \tag{10.3}$$

as u, λ_1, λ_2 approach zero. Our objective is to generalize the discussion in Section 9 for a scalar equation. More precisely, we will impose natural hypotheses on the function C, the matrix L and constant vector k which will permit the application of the procedure of Section 9. With these hypotheses, the complete description of the bifurcation is obtained. Details of the proofs will not be given and can be found in Chow, Hale and Mallet-Paret [17].

Our first hypothesis concerns the cubic function $C(u)$ and ensures that the cubic terms in the components of the vector C are important.

(H1) $C(u) = 0$ implies $u = 0$.

Lemma 10.1 If (H1) is satisfied, there is a neighborhood V of $(u,\lambda) = (0,0)$ and a constant $\beta \neq 0$ such that any solution of Equation (10.1) in V must satisfy

$$|u| \leqslant \beta\,(|\lambda_1|^{1/2} + |\lambda_2|^{1/3}).$$

Proof. Proceed exactly as in the proof of Lemma 9.1 to obtain a contradiction by choosing a subsequence so that $u_n/|u_n|$ approaches a constant vector γ_0 whose magnitude is obviously one and $C(\gamma_0) \neq 0$.

As for the scalar case, this justifies certain scalings in the variables. If

$$u = \lambda_2^{1/3} v, \quad \lambda_1 = \lambda_2^{2/3} \mu, \tag{10.4}$$

then the equivalent equations for v are

$$C(v) + \mu L v + k = O(|\lambda_2|^{1/3}). \tag{10.5}$$

In the scalar case, the solutions of $C(v) + k = 0$ were simple and therefore bifurcations could not occur close to the λ_2 axis in the sense of the parametrization in Relation (10.4). We certainly want the same situation here so we make the hypothesis:

(H2) If $C(v) + k = 0$ then $\det \partial C(v)/\partial v \neq 0$.

Lemma 10.2 Hypothesis (H2) implies there is a $\delta > 0$ such that all solutions of Equation (10.5) are simple for $|\mu| < \delta$, $|\lambda_2| < \delta$.

Proof. This is a direct application of the implicit function theorem.

Lemmas 10.1, 10.2 imply that it is legitimate to reparametrize a neighborhood of $\lambda = 0$ and the variable u by the relation

$$u = |\lambda_1|^{1/2} v, \quad \lambda_2 = |\lambda_1|^{3/2} \nu, \tag{10.6}$$

and only consider λ_1 small and ν in a bounded set. The resulting equivalent equations are

$$G(v,\nu,\lambda_1) \stackrel{\text{def}}{=} C(v) \pm Lv + \nu k - O(|\lambda_1|^{1/2}) \tag{10.7}$$

where $\pm$ designates the sign of λ_1 as before.

To formulate the next hypothesis, let

$$\begin{aligned} g(v,\nu) &= C(v) \pm Lv + \nu k, \\ \Delta(v) &= \det [\partial C(v)/\partial v \pm L], \\ \Delta_1(v,\nu) &= \det [\partial(g,\Delta)/\partial(v,\nu)]. \end{aligned} \tag{10.8}$$

The bifurcation curves must be determined from values of ν, λ_1 for which there are multiple solutions of Equation (10.7). In particular, for $\lambda_1 = 0$, the equations

$$g(v,\nu) = 0, \quad \Delta(v,\nu) = 0, \tag{10.9}$$

must be satisfied. These are three equations for the three unknowns (v_1, v_2, ν). Generically, these solutions should be simple (this was the case of one dimension in Section 9) and thus we impose the following hypothesis:

(H3) If $g(v,\nu) = 0$, $\Delta(v,\nu) = 0$, then $\Delta_1(v,\nu) \neq 0$.

The crucial lemma is

<u>Lemma 10.3</u> If (H1) - (H3) is satisfied, there is a $\delta > 0$ such that if (v_0, ν_0) is a solution of Equation (10.9), there is a unique solution

$v^*(\lambda_1)$, $\nu^*(\lambda_1)$, $|\lambda_1| < \delta$, $v^*(0) = v_0$, $\nu^*(0) = \nu_0$ of the equations

$$G(v,\nu,\lambda_1) = 0, \quad \det \partial G(v,\nu,\lambda_1)/\partial v = 0.$$

Furthermore, the curve $\lambda_2 = |\lambda_1|^{3/2}\nu^*(\lambda_1)$ is a bifurcation curve for Equation (10.1) and the number of solutions changes by exactly two as this curve is crossed. The double solution u on the bifurcation curve is given by $u = |\lambda_1|^{1/2}v^*(\lambda_1)$, $|\lambda_1| < \delta$. Finally, all bifurcation curves are obtained in this manner and they are only finite in number.

Proof. The implicit function theorem implies everything except the assertions about the fact that $\nu^*(\lambda_1)$ gives a bifurcation curve and the number of solutions changes by two as the curve is crossed. The proof of this fact is similar to the one in Section 9. Only the ideas are given and the computations may be found in [17]. For $\lambda_1 = 0$, the equation $g(u,\nu) = 0$ represents the intersection of two cubics. One can show that Hypothesis (H3) implies that when these two cubics intersect and are tangent, then the contact is of second order and, furthermore, varying the parameter ν moves the cubics apart as shown in Figure 8.

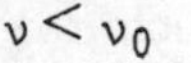

$\nu < \nu_0$ $\qquad$ $\nu = \nu_0$ $\qquad$ $\nu > \nu_0$

Figure 8

Lemma 10.3 gives a complete picture of the bifurcations for Equation (10.1). For a particular example, the verification of the hypotheses and the computation of the approximate bifurcation curves and solutions may be effectively accomplished on a computer. The bifurcation curves are given approximately by the cusps $\lambda_2 = |\lambda_1|^{3/2}\nu_0$ and the approximate solutions are given by $u = |\lambda_1|^{1/2}v_0$ where (v_0, ν_0) satisfy the equations (10.9).

The hypotheses given above are best possible in a certain sense which will be discussed in a later section.

11 BIFURCATION IN TWO DIMENSIONS. QUADRATIC NONLINEARITIES AND TWO PARAMETERS

The analysis in Section 10 is easily adapted to the solution of Equation (10.1) when

$$f(u,\lambda) = Q(u) + \lambda_1 Lu + \lambda_2 k + \text{h.o.t.}, \tag{11.1}$$

where L,k are the same as in (10.3) and

$$\begin{aligned} &Q: \mathbb{R}^2 \to \mathbb{R}^2 \text{ is a homogeneous quadratic,} \\ &\text{h.o.t.} = O(|u|^3 + |\lambda_1 u|^2 + |\lambda_1|^2 + |\lambda_1\lambda_2| + |\lambda_2|^2 + |\lambda_2 u|) \end{aligned} \tag{11.2}$$

as u, λ_1, $\lambda_2 \to 0$. The Hypotheses (H1) - (H3) are modified to

(H1') $Q(u) = 0$ implies $u = 0$.

With Hypothesis (H2'), the a priori bounds on the solutions are

$$|u| \leqslant \beta(|\lambda_1| + |\lambda_2|^{1/2}). \tag{11.3}$$

If $u = |\lambda_2|^{1/2} v$, $\lambda_1 = |\lambda_2|^{1/2}\mu$, the equivalent equations are

$$Q(v) + \mu Lv \pm k = O(|\lambda_2|^{1/2}),$$

and Hypothesis (H2) is replaced by

(H2') If $Q(v) \pm k = 0$, then $\det \partial Q(v)/\partial v \neq 0$.

With the scaling

$$u = \lambda_1 v, \quad \lambda_2 = \lambda_1^2 \nu, \tag{11.4}$$

the equivalent equations are

$$Q(v) + Lv + \nu k = O(|\lambda_1|). \tag{11.5}$$

If we define

$$h(v,\nu) = Q(v) + Lv + \nu k,$$
$$\tilde{\Delta}(v) = \det(\partial Q(v)/\partial v + L).$$
$$\tilde{\Delta}_1(v,\nu) = \det \partial(h,\tilde{\Delta})/\partial(v,\nu),$$

then Hypothesis (H3) is replaced by

(H3') If $h(v,\nu) = 0$, $\tilde{\Delta}(v) = 0$, then $\tilde{\Delta}_1(v,\nu) \neq 0$.

The conclusions are the same as before and the approximate bifurcation curves are obtained from the solutions (finite in number) of the equations

$$h(v,\nu) = 0, \quad \tilde{\Delta}(\nu) = 0, \tag{11.6}$$

and the scaling (11.4).

12 APPLICATIONS TO A SPECIAL FAMILY IN BANACH SPACE

In this section, we discuss the implications of the results in Sections 10 and 11 for the family of mappings given by Equation (5.7); that is

$$M(x,\lambda) = x-(a+\lambda_1)F(x) - \lambda_2 G(x), \tag{12.1}$$

where F,G have continuous derivatives up through order three,

$$\begin{aligned} &a \neq 0, \quad F(0) = 0, \\ &\dim \mathfrak{N}(I-aF'(0)) = 2 = \operatorname{codim} \mathcal{R}(I-aF'(0)). \end{aligned} \tag{12.2}$$

Let $w = (w_1, w_2)$ be a basis for $\mathfrak{N}(I-aF'(0))$ and $z = (z_1, z_2)$ be a basis for a complementary subspace of $\mathcal{R}(I-aF'(0))$. If $u = (u_1, u_2) \in \mathbb{R}^2$, then the bifurcation functions $f = (f_1, f_2) \in \mathbb{R}^2$ are determined from Equation (2.3) by the relation

$$\begin{aligned} z\cdot f(u,\lambda) = (I-E)[\, &a(F(w\cdot u + y^*(w\cdot u,\lambda)) - F'(0)\, y^*(w\cdot u,\lambda) \\ &+ \lambda_1 F(w\cdot u + y^*(w\cdot u,\lambda)) + \lambda_2 G(w\cdot u + y^*(w\cdot u,\lambda))], \end{aligned}$$

where $z\cdot f = z_1 f_1 + z_2 f_2$ and $w\cdot u = w_1 u_1 + w_2 u_2$.

If $f(u,\lambda)$ satisfies either relation (10.2) or (11.1), then

$$z\cdot Lu = (I-E)F'(0)w\cdot u = \frac{1}{a}(I-E)w\cdot u, \tag{12.3}$$

$$z\cdot k = (I-E)G(0), \tag{12.4}$$

$$z\cdot Q(u) = (I-E)F''(0)(w\cdot u, w\cdot u), \tag{12.5}$$

$$z\cdot C(u) = (I-E)F'''(0)(w\cdot u, w\cdot u, w\cdot u). \tag{12.6}$$

Therefore, the hypotheses (H1-H3), (H1'-H3') may be interpreted directly in terms of the functions F,G. If either of these sets of hypotheses is satisfied, then the bifurcations for the equation

$$x - (a + \lambda_1)F(x) - \lambda_2 G(x) = 0,$$

are determined by the methods of Sections 10, 11.

13 APPLICATIONS TO RECTANGULAR PLATES

In this section, we apply the results of the previous sections to the buckling of a simply supported rectangular plate subject to a lateral force and normal loading when the first eigenvalue has multiplicity two; that is, the length of one edge of the plate is $\sqrt{2}$ times the length of the other. The von Kármán equations and boundary conditions are given by Equations (6.1), (6.2) and the abstract equations are given by Equation 6.8 with $\alpha = 0$. The effect of the multiple eigenvalue on the bifurcations has been discussed by numerous authors; see, for example, Bauer, Reiss and Keller [6], Bauer and Reiss [7], Greenlee [27], Keener [41,42], Kirchgassner [44], Knightly and Sather [45], Krasnoselskii [51-53], McLeod and Sattinger [59], Matkowsky and Putnik [65], Pimbley [68] and Stakgold [78].

If $\ell = \sqrt{2}$, the first eigenvalue λ_0^{-1} of the linear operator L is $\lambda_0 = 9\pi^2/2$, $\dim \mathfrak{N}(I-\lambda_0 L) = 2 = \operatorname{codim} \mathcal{R}(I-\lambda_0 L)$ and

$$\begin{aligned}\phi_{11}(x,y) &= \frac{2^{7/4}}{3\pi^2} \sin \frac{\pi x}{\sqrt{2}} \sin \pi y, \\ \phi_{21}(x,y) &= \frac{2^{7/4}}{6\pi^2} \sin \frac{2\pi x}{\sqrt{2}} \sin \pi y,\end{aligned} \qquad (13.1)$$

are basis vectors for $\mathfrak{N}(I-\lambda_0 L)$ as well as coker $\mathcal{R}(I-\lambda_0 L)$.

If $w = u_1\phi_{11} + u_2\phi_{21} \in \mathfrak{N}(I-\lambda_0 L)$, $u = (u_1,u_2) \in \mathbb{R}^2$, then by an application of the Liapunov-Schmidt procedure and the formulas of the previous section, the bifurcation function $f = (f_1,f_2)$ and bifurcation equations are given by

$$f_1 \overset{\text{def}}{=} au_1^3 + bu_1u_2^2 - \lambda_1 u_1 - \lambda_2 k_1 + \text{h.o.t.} = 0,$$
$$f_2 \overset{\text{def}}{=} bu_1^2u_2 + cu_2^3 - \lambda_1 u_2 - \lambda_2 k_2 + \text{h.o.t.} = 0, \tag{13.2}$$

where h.o.t. $= O(|u|^4 + |\lambda_1^2 u| + |\lambda_1 u|^2 + \lambda_2^2 + |\lambda_2 u|)$ as λ_1, λ_2, $u \longrightarrow 0$, and

$$\lambda_1 = \frac{\lambda}{\lambda_0} - 1, \quad \lambda_2 = \nu,$$

$$a = \tfrac{1}{2}\| B(\phi_{11}, \phi_{11})\|_X^2 \cong 3.945001 \times 10^{-4},$$

$$b = \| B(\phi_{11}, \phi_{21})\|_X^2 + \tfrac{1}{2} \langle B(\phi_{11}, \phi_{11}), B(\phi_{21}, \phi_{21})\rangle_X \cong 5.007428 \times 10^{-4}, \tag{13.3}$$

$$c = \tfrac{1}{2}\| B(\phi_{21}, \phi_{21})\|_X^2 \cong 1.623543 \times 10^{-4},$$

$$k_i = \langle q, \phi_{i1}\rangle, \quad i = 1,2.$$

The computation of the explicit values of these constants is not trivial and requires the evaluation of Δ^{-2} which can be obtained from infinite series (see [17], [65]).

Equations (13.2) have the form (10.2) and, therefore, it remains to verify Hypotheses (H1-H3) to obtain the complete bifurcation picture. For a loading function q such that $k_1 = k_2 = 1$, a complete analysis of these hypotheses was given. The restriction $k_1 = k_2 = 1$ is not essential to the analysis but does represent generically the nature of the bifurcations. We summarize the results of this analysis which was accomplished with the aid of a computer.

In presenting the results of the calculations, we say a solution u of Equation (13.2) is a stable node $(--)$ if the two eigenvalues of the linear variational equation are negative, u is a saddle $(+-)$ if one eigenvalue is negative and one is positive and an unstable node $(++)$ if both eigenvalues are positive. A saddle will also be called unstable.

As remarked in Section 10, the approximate bifurcation curves and solutions on the bifurcation curves are given by

$$\lambda_2 = |\lambda_1|^{3/2}\nu_0, \quad u = |\lambda_1|^{1/2} v_0, \tag{13.4}$$

where (v_0, ν_0) satisfy the equations

$$\begin{aligned} &g_1(v_1, v_2, \nu) = 0, \quad g_2(v_1, v_2, \nu) = 0, \\ &\Delta_1(v_1, v_2, \nu) = \det \frac{\partial(g_1, g_2)}{\partial(v_1, v_2)} = 0. \end{aligned} \tag{13.5}$$

The Hypotheses (H1) - (H3) were verified for this example and the results of the calculations are given in Table 13.1.

ν_0	v_{01}	v_{02}
-15.936	-17.088	60.245
-13.610	21.326	19.085
- 4.0493	-48.957	-16.550
- 1.829	-47.370	12.334
1.829	47.370	-12.334
4.0493	48.957	16.550
13.610	-21.326	-19.085
15.936	17.088	-60.245

Second order contacts in the von Kármán equations with loading $k_1 = 1$, $k_2 = 1$.

Table 13.1

The bifurcation curves are shown in Figure 9, consisting of eight arcs with the number of stable and unstable solutions in any region of (λ_1, λ_2)-space as designated in the figure. Typical solution versus λ_1 plots with λ_2 fixed are shown in Figure 10.

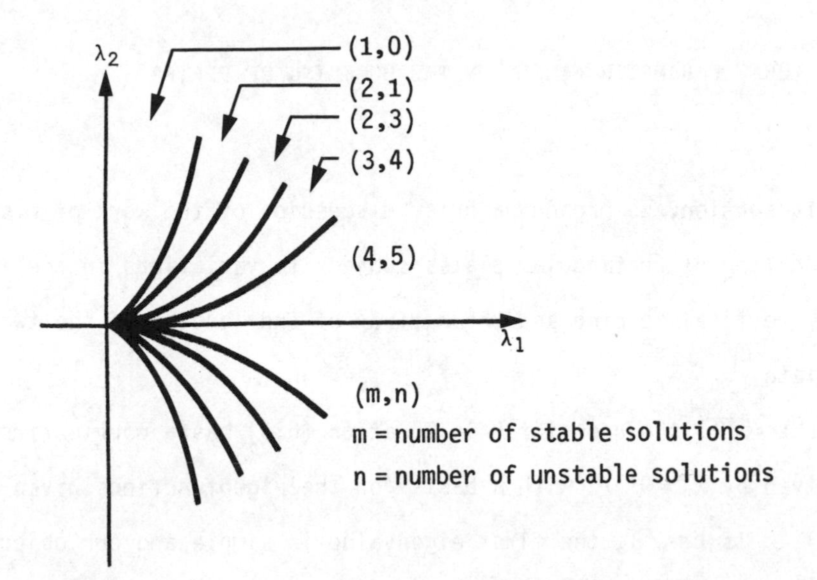

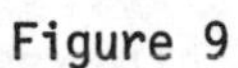
Figure 9

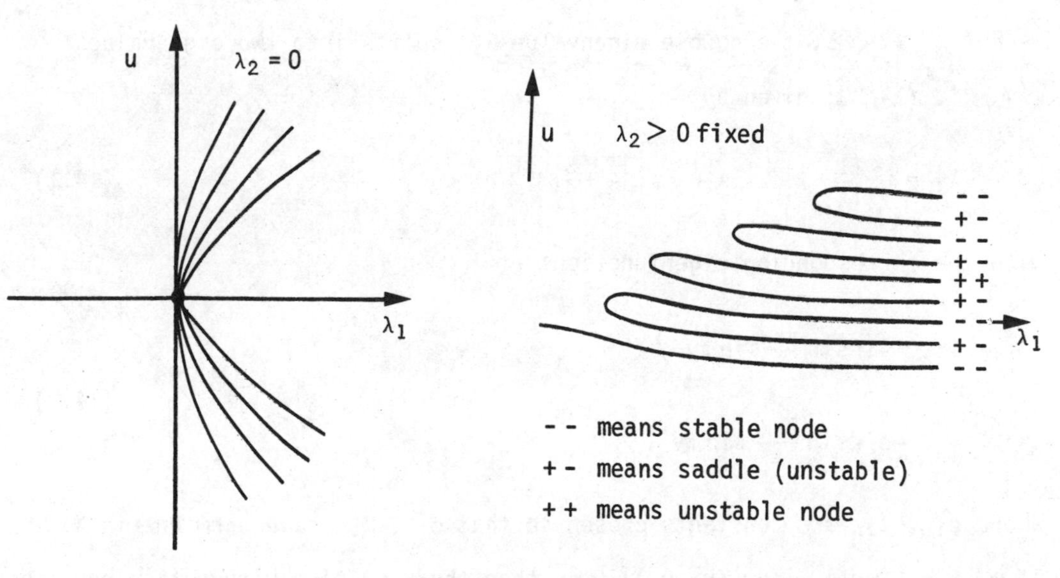

Figure 10

14 A THREE PARAMETER FAMILY IN THE BUCKLING OF PLATES

In this section, we present a brief discussion of the work of List [56] on the buckling of rectangular plates subject to variations in the lateral force, vertical loading and the ratio ℓ of the lengths of the two edges of the plate.

For $\ell = \sqrt{2}$, the operator L in Equation (6.3) has a double first eigenvalue λ_0^{-1} given by $\lambda_0 = 9\pi^2/2$ with a basis for the eigenfunctions given by Relation (13.1). If $\ell \neq \sqrt{2}$, the first eigenvalue is simple and our objective is to study the effect on bifurcation of the coalesence of eigenvalues of L. As we will see, secondary bifurcations (i.e., bifurcations from a bifurcation curve) can occur, a phenomena which has also been discussed by Bauer, Keller and Reiss [6], Bauer and Reiss [7] and Keener [41].

For $\ell \neq \sqrt{2}$, the double eigenvalue λ_0^{-1} splits into two eigenvalues $(\lambda_{11})^{-1}$, $(\lambda_{21})^{-1}$ given by

$$\lambda_{11} = \pi^2 \left(\frac{1}{\ell} + \ell\right)^2, \quad \lambda_{21} = \pi^2 \left(\frac{2}{\ell} + \frac{\ell}{2}\right)^2, \tag{14.1}$$

with the corresponding eigenfunctions

$$\begin{aligned} \phi_{11} &= c_{11} \sin \frac{\pi x}{\ell} \sin \pi y , \\ \phi_{21} &= c_{21} \sin \frac{2\pi x}{\ell} \sin \pi y , \end{aligned} \tag{14.2}$$

where c_{11}, c_{21} are constants chosen so that ϕ_{11}, ϕ_{21} have norm one in X. If $w = u_1\phi_{11} + u_2\phi_{21}$, $u = (u_1, u_2) \in \mathbb{R}^2$, then the resulting bifurcation equations are given by

$$f_1 \overset{\text{def}}{=} -au_1^3 - bu_1u_2^2 + \lambda_1 u_1 + \lambda_3 k_1 = \text{h.o.t.},$$
$$f_2 \overset{\text{def}}{=} -bu_1^2u_2 - cu_2^3 + \lambda_2 u_2 + \lambda_3 k_2 = \text{h.o.t.}, \tag{14.3}$$

where a,b,c,k_1,k_2 are the constants given in Relation (13.3) and

$$\lambda_1 = \left(\frac{\lambda}{\lambda_{11}} - 1\right), \quad \lambda_2 = \left(\frac{\lambda}{\lambda_{21}} - 1\right), \quad \lambda_3 = \nu, \tag{14.4}$$

$$\text{h.o.t.} = O(|u|^4 + (\lambda_1^2 + \lambda_2^2)|u| + (|\lambda_1| + |\lambda_2|)|u|^2\lambda_3^2 + |\lambda_3 u|.$$

In the remaining discussion, we set $\lambda = (\lambda_1,\lambda_2,\lambda_3)$ even though we have used λ as the lateral force before. No confusion should arise.

Equations (14.3) involve three parameters and reduce to the Equations (13.2) in two parameters when $\ell = \sqrt{2}$. To determine the nature of the bifurcations, the theory of Section 10 is first generalized abstractly and then applied to the specific Equations (14.3). We do not present the complete details of the theory but only state the results and implications. Suppose Equations (14.3) are written as

$$f(u,\lambda) = C(u) + \lambda_1 L_1 u + \lambda_2 L_2 u + \lambda_3 k = \text{h.o.t.} \tag{14.5}$$

where $f \in \mathbb{R}^2$, $u \in \mathbb{R}^2$, $\lambda \in \mathbb{R}^3$, and h.o.t. is defined in Relation (14.4), $C(u)$ is a homogeneous cubic in $\mathbb{R}^2$, L_1, L_2 are 2×2 matrices and $k \in \mathbb{R}^2$ is given. For any value of the parameters λ for which the corresponding hypotheses (H1 - H3) of Section 10 are satisfied, the conclusions of that section apply and the bifurcations are known. With three parameters entering as in Equation (14.5), it is possible for Hypothesis (H3) to be violated at some value of λ. The difficulty in the analysis is to discover the appropriate additional hypothesis which will permit a complete description of the bifurcations. To state the hypotheses, some additional notation is needed.

Let $v=(v_1,v_2)\in\mathbb{R}^2$, $\mu=(\mu_1,\mu_2)\in\mathbb{R}^2$, $g=(g_1,g_2)\in\mathbb{R}^2$,

$$\begin{aligned}
g(v,\mu) &= C(v)\pm L_1u+\mu_1L_2v+\mu_2k,\\
\tilde{g}(v,\mu_1) &= k_2g_1(v,\mu)-k_1g_2(v,\mu),\\
\Delta_1(v,\mu_1) &= \det(\partial g/\partial v),\\
\Delta_2(v,\mu_1) &= \det\frac{\partial(g,\Delta_1)}{\partial(v,\mu_2)}\\
\Delta^*(v,\mu_1) &= \det\frac{\partial(\tilde{g},\Delta_1)}{\partial(v_1,v_2)},\\
\Delta^0(v,\mu_1) &= \det\frac{\partial(\tilde{g},\Delta_1,\Delta^*)}{\partial(v_1,v_2,\mu_1)},
\end{aligned}\tag{14.6}$$

where the ± is obtained via a scaling as in Section 10 using $u=|\lambda_1|^{1/2}v$, $\lambda_2=\mu_1\lambda_1$, $\lambda_3=\mu_2|\lambda_1|^{3/2}$ with the + sign denoting λ_1 is positive and the - sign denoting λ_1 is negative.

The hypotheses imposed are the following:

(J1) $C(u)=0$ implies $u=0$.

(J2) If $C(v)+k=0$ then $\det\partial C(v)/\partial v\neq 0$.

(J3) If $\tilde{g}(v,\mu_1)=0$, $\Delta_1(v,\mu_1)=0$, $\Delta^*(v,\mu_1)=0$, then $\Delta^0(v,\mu_1)\neq 0$.

(J4) If $g(v,\mu)=0$, $g(\tilde{v},\mu)=0$, $\Delta_1(v,\mu_1)=0$, $\Delta_2(\tilde{v},\mu_1)=0$, then

$$\det\frac{\partial(g(v,\mu),g(\tilde{v},\mu),\ \Delta_1(v,\mu_1),\ \Delta_2(\tilde{v},\mu_1))}{\partial(v,\tilde{v},\mu)}\neq 0.$$

Hypotheses (J1), (J2) are the same as Hypotheses (H1), (H2) in Section 10. Hypotheses (J3), (J4) are the generic hypotheses replacing Hypothesis (H3).

We now summarize the results of List [56] without proofs. For the statement of the results, let $\eta=(\eta_1,\eta_2)\in\mathbb{R}^2$, $m\in\mathbb{R}$, $\nu\in\mathbb{R}$ and consider the equations

$$g(\eta,m,\nu),\quad \Delta_1(\eta,m,\nu)=0. \tag{14.7}$$

Theorem 14.1 For all but a finite number of values m, any solution $\eta^*(m)$, $\nu^*(m)$ of Equation (14.7) satisfies $\Delta^*(\eta^*,m) \neq 0$ and Hypothesis (H3). Thus results of Section 10 are immediately applicable. Hence, excepting finitely many m $(-1 \leq m \leq 1)$, the half-plane $P_m = \{\lambda_1,\lambda_2,\lambda_3): \lambda_1 > 0, \frac{\lambda_2}{\lambda_1} = m\}$ satisfies the following:

There exist a finite collection of curves

$$A_{\nu^*(m)}: \lambda_3 \sim \nu^*(m)\lambda_1^{3/2}, \quad \lambda_1 \to 0^+ \tag{14.8}$$

on P_m, with bifurcation occurring at

$$u \sim \eta^*(m)\lambda_1^{1/2}, \quad \lambda_1 \to 0^+,$$

where $\eta^*(m)$, $\nu^*(m)$ satisfy Equations (14.7). In crossing one of these curves $A_{\nu^*(m)}$, moving upward (λ_3 increases) in the plane P_m, the number of solutions to Equation (14.3) increases or decreases by two, depending on whether $\Delta_2(\eta^*(m),m)$ is positive or negative, respectively. One solution is an unstable saddle and the other is a node, stable or unstable, depending on whether the quantity $\lambda^* = (\text{trace } \Delta_1(\eta^*,m))$ is negative or positive. In each region between the curves defined by Relation (14.8) the number of solutions of Equation (14.3) remains constant.

For the statement of the next result we need the equations

$$\tilde{g}(\eta,m) = 0, \quad \Delta_1(\eta,m) = 0. \tag{14.9}$$

Theorem 14.2 At those values $\overline{m}$ for which there exist $(\nu^*(\overline{m}),\eta^*(\overline{m}))$ satisfying both Equations (14.9) and the condition $(\Delta^*(\eta^*,\overline{m}) = 0)$, two distinct values of $\nu^*(m)$ coalesce and vanish. If $\Delta^0(\eta^*,\overline{m}) > 0$, these values vanish for $m < \overline{m}$; if $\Delta^0(\eta^*,\overline{m}) < 0$, they vanish for $m > \overline{m}$. In either case, the distinct bifurcation generating arcs $A_{\nu_1^*(m)}$, $A_{\nu_2^*(m)}$ $(\nu_1^*(m) \neq \nu_2^*(m))$

coalesce into arcs $A_{\nu_1*(\overline{m})}$, $A_{\nu_2*(\overline{m})}$ of identical asymptotic behavior ($\nu_1*(\overline{m}) = \nu_2*(\overline{m})$) and then vanish. Hence, two distinct bifurcation sheets, generated by the arcs $A_{\nu_1*(m)}$ and $A_{\nu_2*(m)}$, coalesce along a curve

$$\gamma = (\hat{m}(\mu),\mu,\hat{\nu}(\mu)),$$

where $\hat{m}(0) = \overline{m}$ and $\hat{\nu}(0) = \nu*(\overline{m})$.

Those values $m = \overline{m}$ for which two distinct solution values η^1,η^2 correspond to a single parameter value $\nu*$ may be obtained by solving the system of equations

$$\begin{aligned} &g(\eta^1,0,\overline{m},\nu*) = 0,\\ &g(\eta^2,0,\overline{m},\nu*) = 0,\\ &\Delta_1(\eta^1,\overline{m}) = 0,\\ &\Delta_1(\eta^2,\overline{m}) = 0. \end{aligned} \tag{14.10}$$

Values $\overline{m}$ for which there exist $(\eta^1,\eta^2,\nu*)$ satisfying Equations (14.10) represent intersections of distinct bifurcation sheets. The hypothesis (J4) merely ensures that such intersections are transversal. Also, the determinant condition in (J4) guarantees, by the implicit function theorem, the local existence of the actual curve

$$\psi:\ (\hat{\hat{m}}(\mu),\mu,\hat{\hat{\nu}}(\mu)) \tag{14.11}$$

of intersection, where $\hat{\hat{m}}(0) = \overline{m}$, $\hat{\hat{\nu}}(0) = \nu*(\overline{m})$ and $\hat{\hat{\nu}}(\mu) = \gamma_{\hat{\hat{m}}(\mu)}(\mu)$, where γ denotes the bifurcation arc (14.8). In crossing such an intersection curve ψ, it is possible for the number of solutions of Equation (14.3) to increase (decrease) by four.

These comments are summarized in:

Theorem 14.3 Each value of $m=\overline{m}$ for which there exists (η^1,η^2,ν^*) satisfying Equation (14.10) corresponds to the transversal intersection of two distinct bifurcation sheets, with intersection occurring along the curve ψ in Equation (14.11). In crossing the curve ψ, it is possible for the number of solutions of Equation (14.3) to increase (decrease) by four.

The theorems of this chapter, then, provide a generic characterization of the bifurcation diagram of the three-parameter equation (14.3). Generically, this diagram consists of a finite collection of sheets emanating from, and encircling, the origin. Pairs of these sheets might coalesce and vanish, or might intersect each other transversally, in the manner we have described.

Let a neighborhood U of zero in the $(\lambda_1,\lambda_2,\lambda_3)$-space be divided into the four regions:

$$\mathcal{R}_{++}=\{(\lambda_1,\lambda_2,\lambda_3):\ \lambda_1>0,\ |\lambda_2|\geqslant\lambda_1\},$$
$$\mathcal{R}_{+-}=\{(\lambda_1,\lambda_2,\lambda_3):\ \lambda_2>0,\ |\lambda_1|\leqslant\lambda_2\},$$
$$\mathcal{R}_{--}=\{(\lambda_1,\lambda_2,\lambda_3):\ \lambda_1<0,\ |\lambda_2|\leqslant|\lambda_1|\},$$
$$\mathcal{R}_{-+}=\{(\lambda_1,\lambda_2,\lambda_3):\ \lambda_2\leqslant 0,\ |\lambda_1|\leqslant\lambda_2\}.$$

In depicting the results for the von Kármán equations, it is convenient to use the m,ν defined by $\lambda_1=m\lambda_2$, $\lambda_3=\nu|\lambda_1|^{3/2}$. Also the computations were made for $k_1=k_2=1$.

As m varies through each of the four regions $\mathcal{R}_{++}$, $\mathcal{R}_{+-}$, $\mathcal{R}_{-+}$, $\mathcal{R}_{--}$, the entire three-parameter bifurcation diagram is constructed. The four sections of Figure 11 contain a complete cross-sectional representation ($\nu^*(m)$ (positive values) plotted versus m) of the bifurcation diagram. Note that the curves of each region extend naturally into those of adjoining regions. Points representing the coalescence of two distinct bifurcation

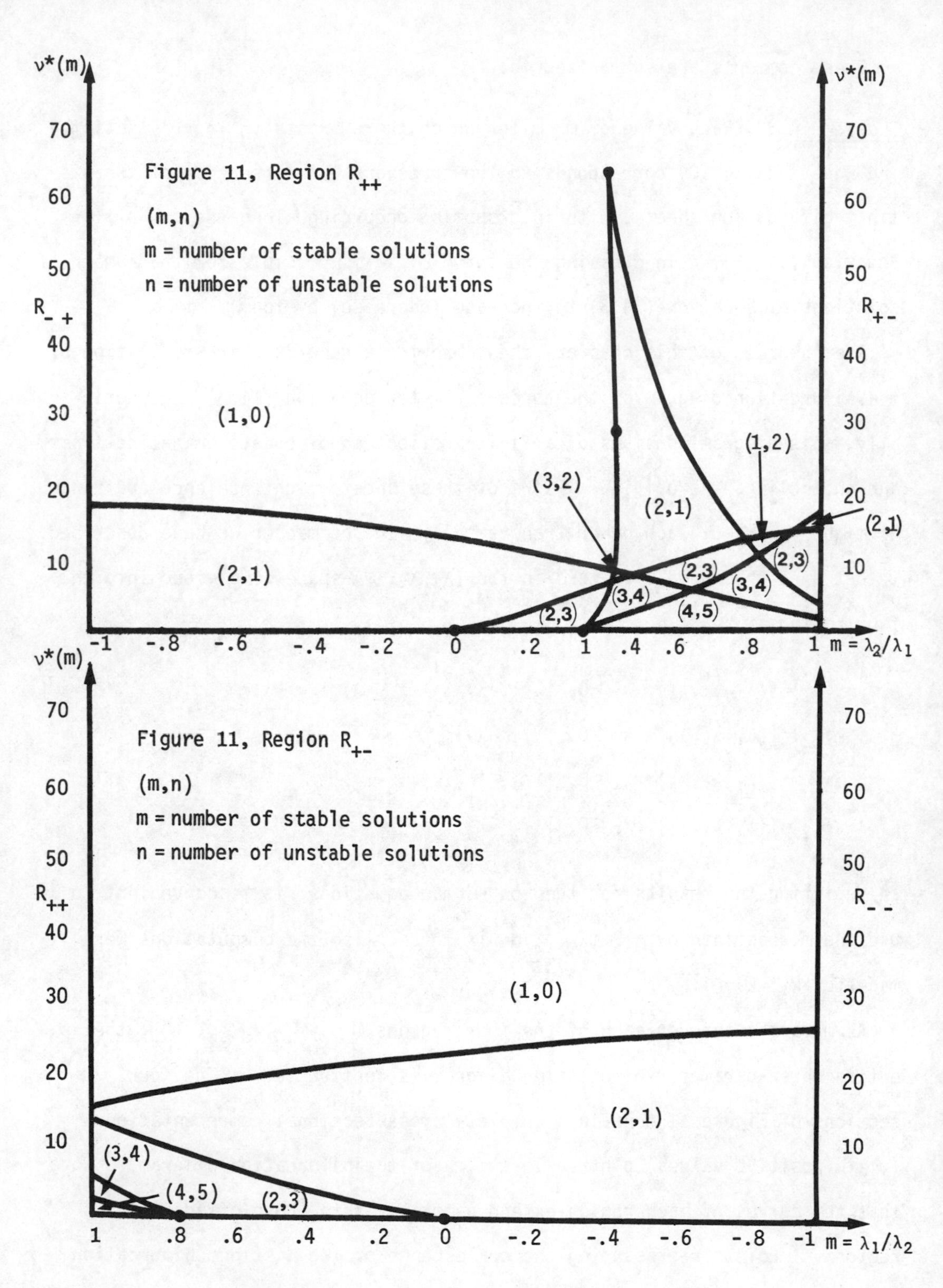
ν*(m)
ν*(m)
70
60
50
40
30
20
10
Figure 11, Region R++
(m,n)
m = number of stable solutions
n = number of unstable solutions
R- +
R+-
(1,0)
(1,2)
(3,2)
(2,1)
(2,1)
(2,1)
(2,3)
(2,3)
(3,4)
(3,4)
(2,3)
(4,5)
-1
-.8
-.6
-.4
-.2
0
.2
.4
.6
.8
1
m = λ2/λ1
ν*(m)
Figure 11, Region R+-
(m,n)
m = number of stable solutions
n = number of unstable solutions
R++
R- -
(1,0)
(2,1)
(3,4)
(4,5)
(2,3)
1
.8
.6
.4
.2
0
-.2
-.4
-.6
-.8
-1
m = λ1/λ2

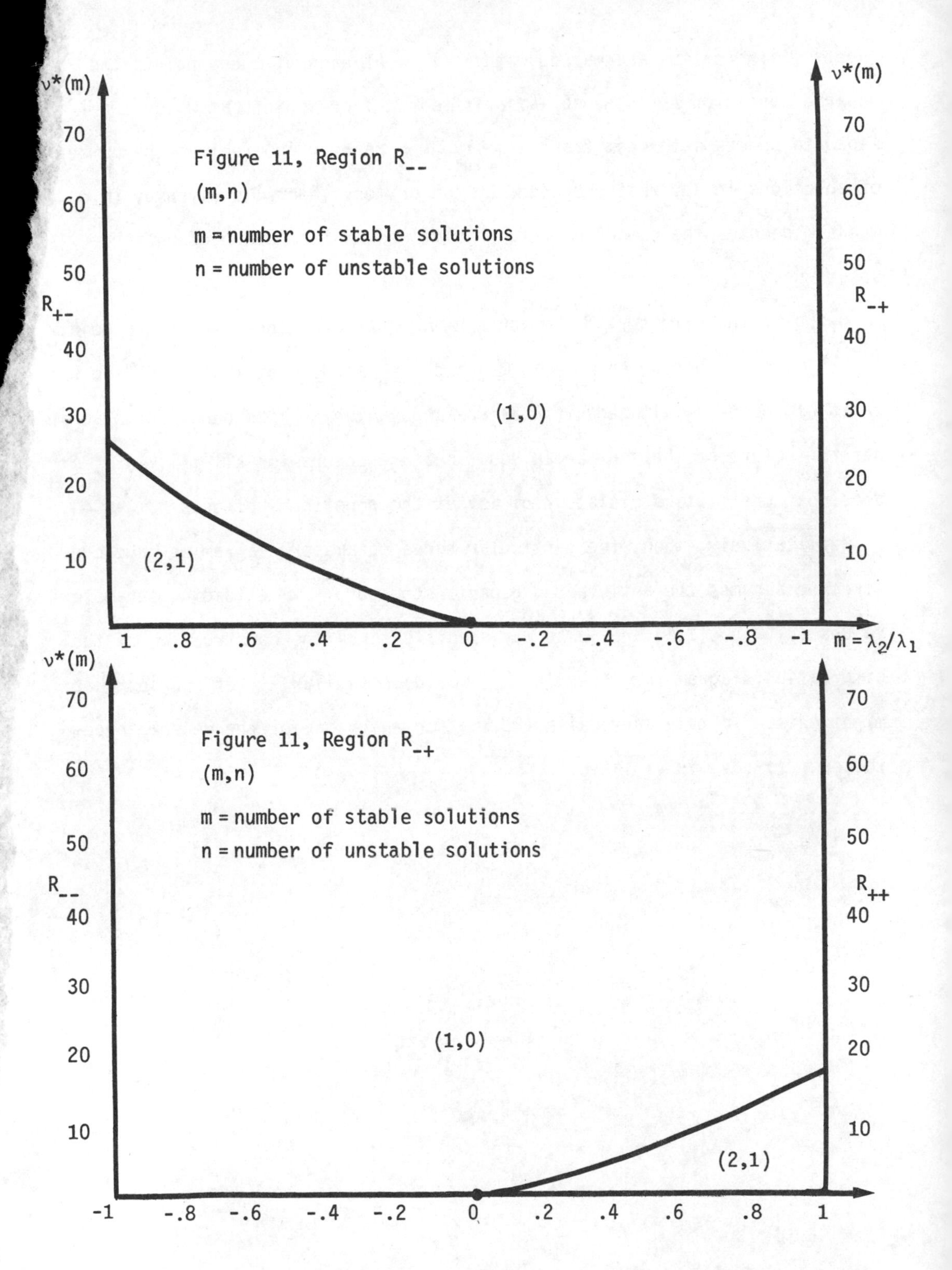
ν*(m)
70
60
50
40
30
20
10
Figure 11, Region R_--
(m,n)
m = number of stable solutions
n = number of unstable solutions
R_+-
R_-+
(1,0)
(2,1)
1 .8 .6 .4 .2 0 -.2 -.4 -.6 -.8 -1
m = λ2/λ1
Figure 11, Region R_-+
(m,n)
m = number of stable solutions
n = number of unstable solutions
R_--
R_++
(1,0)
(2,1)
-1 -.8 -.6 -.4 -.2 0 .2 .4 .6 .8 1

sheets ($\Delta^*(\eta^*,m) = 0$) are marked with a($\cdot$). At each of these points the generic condition $\Delta^0(\eta^*,m) \neq 0$ is satisfied. Those points at which two separate sheets intersect transversally are readily apparent. The number of solutions in any region bounded by two or more sheets is given by (m,n), where m denotes the number of stable solutions, n the number of unstable solutions.

Owing to the symmetry of the von Kármán equations, there exist two points ($\cdot^1,\cdot^2$) (corresponding to $m = \frac{c}{b}$ in $\mathcal{R}_{++}$ and $m = \frac{a}{b}$ in $\mathcal{R}_{+-}$) at which two distinct pairs of sheets simultaneously coalesce and vanish. This behavior is an inherent feature of the von Kármán equations and occurs for all (k_1,k_2). It does not constitute a violation of any of the generic hypotheses (J1)-(J4).

Qualitatively, then, the particular three-parameter diagram we have constructed assumes the anticipated generic structure. The loading parameter values $(k_1,k_2) = (1,1)$ are in no way special, since the construction of the bifurcation diagram for other (k_1,k_2) would be similar. (Of course, there might exist finitely many pairs (k_1,k_2) for which one of the generic hypotheses (J2)-(J4) is violated.

15 BUCKLING OF CYLINDRICAL SHELLS WITH SMALL CURVATURE

Following Mallet-Paret [63], we discuss in this section the effect on buckling of an imperfection caused by a small curvature in a rectangular plate. To make the problem as simple as possible, we suppose only lateral loading and the ratio of the lengths of the edges of the plate are $\sqrt{2}$. From the model given in Section 6, this means $\nu = 0$ and $\ell = \sqrt{2}$. Also, to avoid further complications, we suppose the initial deflection of the plate is αw_0, where

$$w_0(x,y) = \frac{1}{2}\, y^2(\sigma x + \tau) \tag{15.1}$$

and σ,τ are fixed parameters. This implies $p = 0$ in Section 6. Thus, the abstract von Kármán equations are

$$C(w) + \alpha Q(w) + (I - \lambda L + \alpha^2 \Lambda^2)w = 0, \tag{15.2}$$

where the functions are defined in Section 6.

Two problems will be discussed, one for which $\sigma \neq 0$, $\tau \neq 0$, in Equation (15.1) and the other for which $\sigma = 0$, $\tau = 1$; that is,

$$w_0(x,y) = \frac{1}{2}\, y^2, \tag{15.3}$$

and the initial displacement of the plate is independent of x. The abstract von Kármán equations for w_0 given in Equation (15.3) are

$$C(w) + \alpha Q(w) + (I - \lambda L + \alpha^2 L^2)w = 0. \tag{15.4}$$

The problem is to discuss the bifurcations in Equations (15.2) and (15.4) when (λ,α) vary in a full neighborhood of $(\lambda_0,0)$, where λ_0 is the first eigenvalue of L. It has multiplicity two with eigenfunctions given by Equation (13.1). Knightly [47] and Sather [73] have considered this problem for $\alpha \neq 0$ small and fixed and λ varying near λ_0. As we shall see, the qualitative structure of the bifurcations are different for these two cases.

The bifurcation pattern for Equation (15.2) and (15.4) can be significantly different. The effect of σ above can be pronounced. Basically for $\sigma = 0$ and small $|\alpha|$, there is still a double eigenvalue at $\lambda = \lambda_0 + \alpha^2/\lambda_0$ for the operator $I - \lambda L + \alpha^2 L^2$. However, for small $|\sigma| \neq 0$, the double eigenvalue at $\lambda = \lambda_0$ in general splits into two nearby simple eigenvalues of the operator $I - \lambda L + \alpha^2 \Lambda^2$. Although some symmetry destroying phenomena also occur when the modified imperfection is considered, perhaps the most significant is this splitting of the double eigenvalues.

Using the Liapunov-Schmidt procedure and the basis vectors in Equation (13.1), the bifurcation equations for (15.2) are given by

$$\begin{aligned} &f^{⌑}(u,\lambda,\alpha) = 0, \quad f^{⌑} = (f_1^{⌑}, f_2^{⌑}) \in \mathbb{R}^2, \; u = (u_1,u_2) \in \mathbb{R}^2, \\ &f^{⌑}(u,\lambda,\alpha) = -\gamma u + \alpha^2 M u + \alpha q^{⌑}(u) + c(u) + O(|u|^5 + \alpha^4 |u|^3) \end{aligned} \tag{15.5}$$

where $c(u)$ are the same cubic terms as in Section 13.

$$\gamma = -1 + \frac{\lambda}{\lambda_0} - \frac{\alpha^2}{\lambda_0},$$

$$M = \begin{bmatrix} k_2 - \dfrac{\alpha^2}{\lambda_0} & k_3 \\ k_3 & k_4 - \dfrac{\alpha^2}{\lambda_0} \end{bmatrix},$$

$$q^{\#}(u) = \begin{pmatrix} q_1(u) \\ q_2(u) \end{pmatrix}, \tag{15.6}$$

$$q_1^{\#}(u) = \langle \phi_{11}, Q^{\#}(u_1\phi_{11} + u_2\phi_{21}) \rangle$$

$$= -(5k_1u_1^2 + 3k_1u_2^2)(\tau + \frac{\sigma}{2^{1/2}}) + 2k_7u_1u_2\sigma,$$

$$q_2^{\#}(u) = \langle \phi_{21}, Q^{\#}(u_1\phi_{11} + u_2\phi_{21}) \rangle$$

$$= -(6k_1u_1u_2)(\tau + \frac{\sigma}{2^{1/2}}) + (k_7u_1^2 + k_8u_2^2)\sigma,$$

and

$$k_2 = \langle \phi_{11}, \Lambda^2\phi_1 \rangle = \left(\frac{4}{81\,\pi^4}\right)\tau^2 + \left(\frac{2^{5/2}}{81\,\pi^4}\right)\tau\sigma + k_5\sigma^2,$$

$$k_3 = \langle \phi_{11}, \Lambda^2\phi_{21} \rangle = -\left(\frac{5(2)^{11/2}}{729\,\pi^6}\right)\tau\sigma - \left(\frac{160}{729\,\pi^6}\right)\sigma^2,$$

$$k_4 = \langle \phi_{21}, \Lambda^2\phi_{21} \rangle = \left(\frac{4}{81\,\pi^4}\right)\tau^2 + \left(\frac{2^{5/2}}{81\,\pi^4}\right)\tau\sigma + k_6\sigma^2, \tag{15.7}$$

$$k_5 \cong 3.440729 \times 10^{-4},$$

$$k_6 \cong 3.675239 \times 10^{-4},$$

$$k_7 \cong 1.939866 \times 10^{-4},$$

$$k_8 \cong 2.288281 \times 10^{-4}.$$

The symbol on $Q^{\#}(w)$ designates that the quadratic terms in the expression for $Q^{\#}(w)$ are evaluated for the linear operator Λ^2.

The bifurcation equations for Equation (15.4) are

$$\begin{aligned} &f^*(u,\lambda,\alpha) = 0, \; f^* = (f_1^*, f_2^*) \in \mathbb{R}^2, \; u = (u_1,u_2) \in \mathbb{R}^2, \\ &f^*(u,\lambda,\alpha) = -\gamma u + \alpha q^*(u) + c(u) + O(|u|^5 + \alpha^2|u|^3), \end{aligned} \tag{15.8}$$

where the star designates that the quadratic terms are obtained from (15.6) by evaluating the function Q(w) in Section 6 at the linear operator $\Lambda = L$. The quadratic term in (15.8) is given by Knightly and Sather (see [47]) as

$$\begin{aligned} &\langle \phi_{11}, Q(u_1\phi_{11} + u_2\phi_{21}) \rangle = -5k_1u_1^2 - 3k_1u_2^2, \\ &\langle \phi_{21}, Q(u_1\phi_{11} + u_2\phi_{21}) \rangle = -6k_1u_1u_2, \\ &k_1 = \frac{256}{(1215)(2)^{1/4}\pi^6} \cong 1.842922 \times 10^{-4}. \end{aligned} \tag{15.9}$$

The complete analysis of the bifurcation equations is very complicated and makes use of extensive scaling properties as well as the inherent symmetry properties in Equations (15.8). The details may be found in Mallet-Paret [63] and we merely state the results. The generic hypotheses imposed are the following:

(H1) $c(u) = 0$ implies $u = 0$,

(H2) $\mp u + c(u) = 0$ implies $\det(\mp I + \frac{\partial c}{\partial u}) \neq 0$,

(H3*) $\left.\begin{aligned} &-\gamma u + q^*(u) + c(u) = 0 \\ &\Delta_1^*(u,\gamma) = 0, \; u \neq 0 \end{aligned}\right\}$ implies $\Delta_2^*(u,\gamma) \neq 0$.

(H3$^{\sharp}$) $\left.\begin{aligned} &(-\gamma I + M)u + q^{\sharp}(u) + c(u) = 0 \\ &\Delta_1^{\sharp}(u,\gamma) = 0, \; u \neq 0 \end{aligned}\right\}$ implies $\Delta_2^{\sharp}(u,\gamma) \neq 0$,

(H4*) $q^*(u) = 0$ implies $u = 0$,

(H5*) $-u + q^*(u) = 0$ implies $\det(-I + \frac{\partial q^*}{\partial u}) \neq 0$,

(H5$^{\sharp}$) M has distinct eigenvalues. If u^0 is one of the eigenvectors then $\langle u^0, q^{\sharp}(u^0) \rangle \neq 0$.

Under the above hypotheses, the complete bifurcations of Equations (15.5) and (15.8) can be given. The specific values of the constants in Relations (15.7) and (15.9) are not important. In the application to our specific buckling problem, these constants must be considered to permit the hypotheses to be satisfied. This is also where the symmetry properties are used. Mallet-Paret [63] has shown that all of these hypotheses can be verified if σ is sufficiently small. The bifurcation diagrams and solutions versus λ for α fixed are shown in Figures 12 - 15. As one can see from these figures, the small change in the initial curvature of the plate forces a pattern of generic bifurcation in the sense that the number of solutions changes by exactly two with increasing lateral force (except at one point where symmetry gives two simultaneous bifurcations).

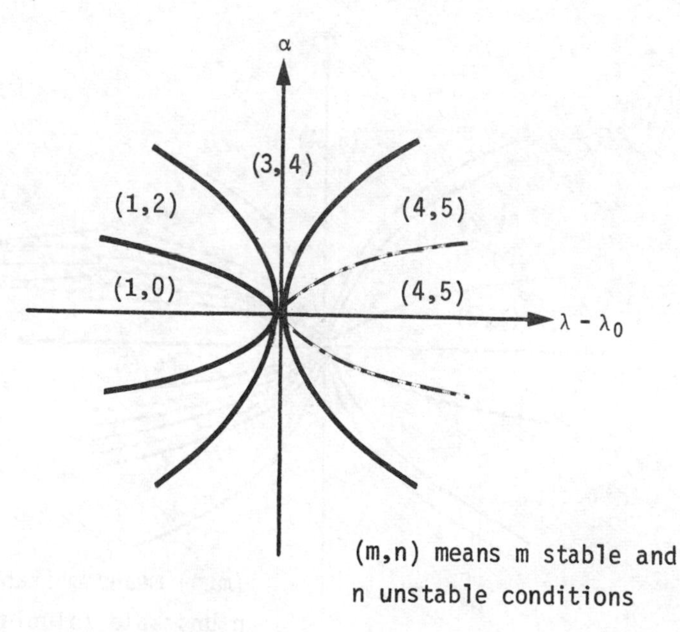

Figure 12 Number of solutions with $w_0(x,y) = \frac{1}{2} y^2$

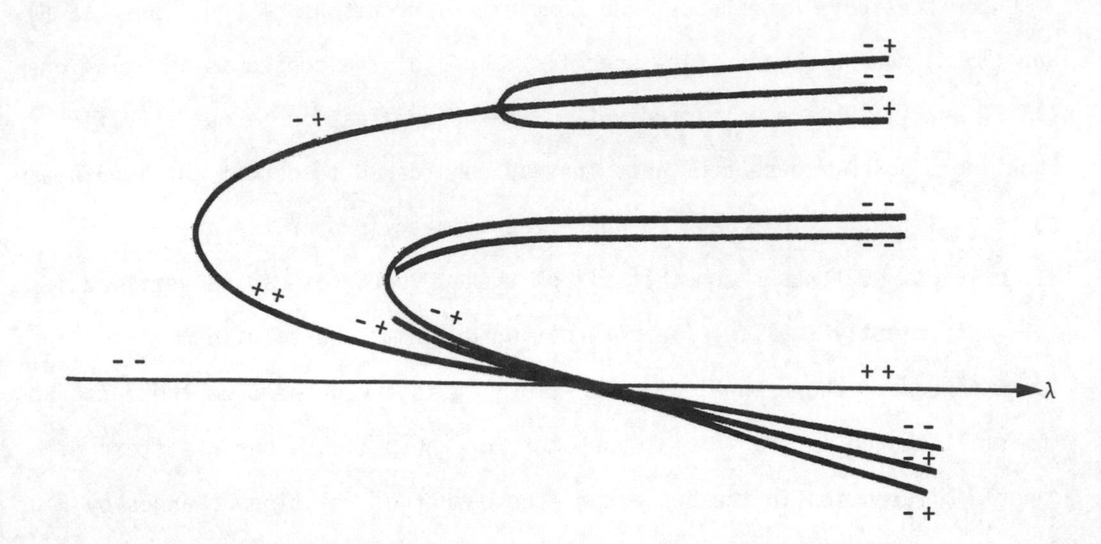

Figure 13 Bifurcation diagram with $w_0(x,y) = \frac{1}{2} y^2$

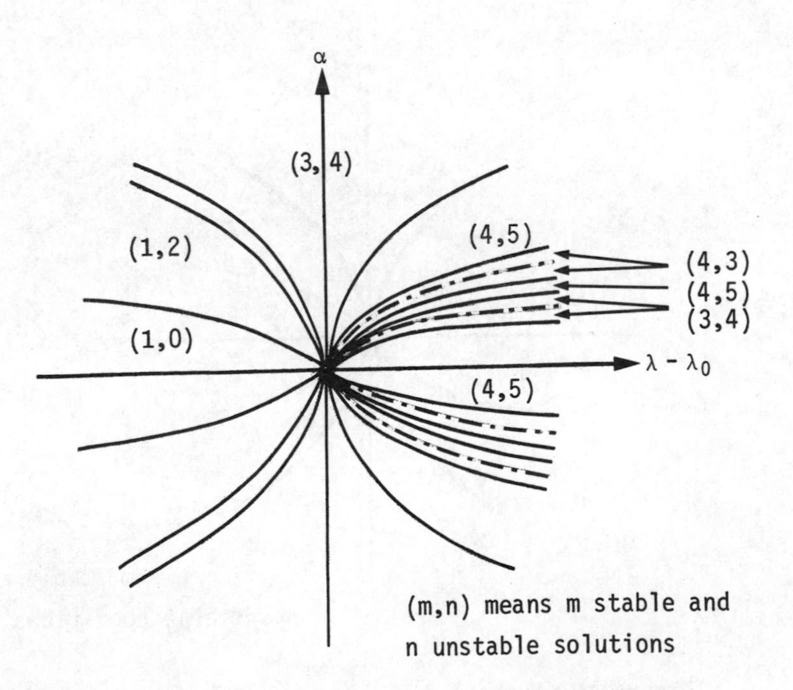

Figure 14 Number of solutions with $w_0(x,y) = \frac{1}{2} y^2 (1 + \sigma x)$, $0 < |\sigma| \ll 1$

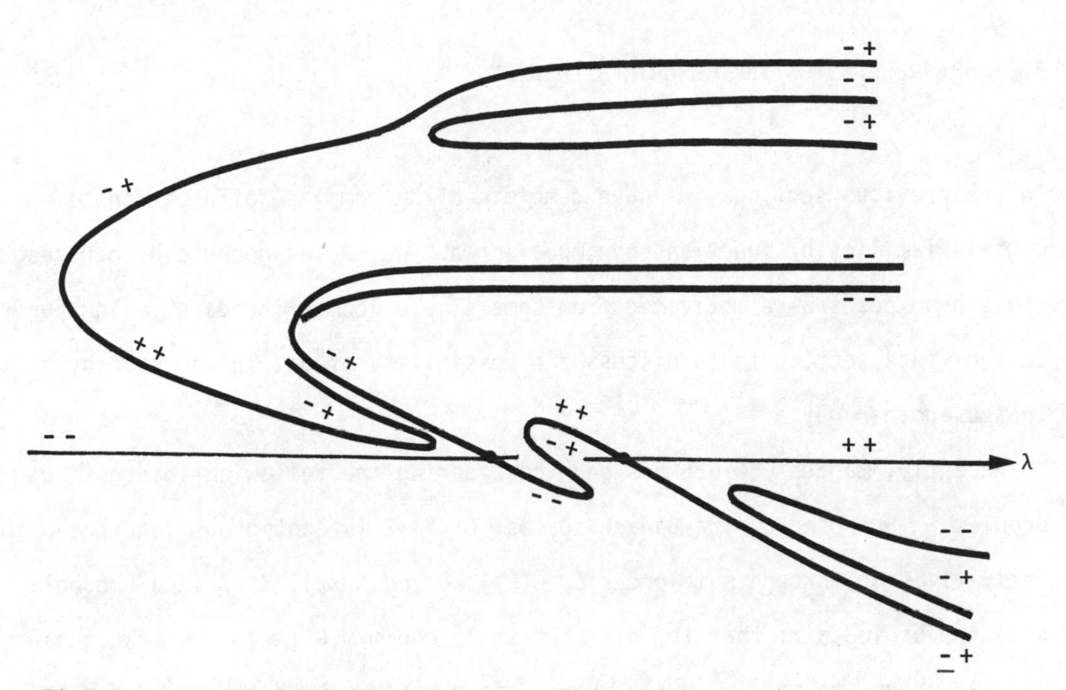

Figure 15 Bifurcation diagram with $w_0(x,y) = \frac{1}{2}y^2(1+\sigma x)$, $0<|\sigma|\ll 1$

16 SOME REMARKS ON THE GENERIC HYPOTHESES

In the previous sections, we have determined the complete bifurcations of certain families of functions by imposing what we called generic hypotheses. These hypotheses were motivated from some simple geometric ideas. The purpose of this section is to discuss the possibility of obtaining these hypotheses another way.

Recently, Magnus [60-62] has been considering the following interesting problem. Suppose Y,Z are Banach spaces, $G: Y \to Z$ is continuous together with derivatives up through some order p, $G(y_0) = 0$ and $G'(x_0): Y \to Z$ is a Fredholm operator of index r; that is, $\dim \mathfrak{N}(G'(x_0)) - \operatorname{codim} \mathcal{R}(G'(x_0)) = r$. What can be said about the set $G^{-1}(0) \cap U$ where U is a neighborhood of x_0? Under the hypothesis $r = 1$, a complete description of $G^{-1}(0) \cap U$ has been given in [61].

These ideas can be applied to bifurcation theory (see Magnus [60] and Shearer [76]). To see how the connection arises, let us consider the family of functions in Section 10:

$$F(u,\lambda,\nu) \stackrel{\text{def}}{=} \frac{1}{6}Cu^3 + \lambda Lu + \nu k + \text{h.o.t.} = 0, \tag{16.1}$$

where $\lambda \in \mathbb{R}$, $\nu \in \mathbb{R}$, h.o.t. is the same order relation as in Equation (10.3), $F \in \mathbb{R}^2$, $u \in \mathbb{R}^2$, L is a 2×2 matrix, $k \in \mathbb{R}^2$ is given and Cuvw denotes a symmetric three linear form on the components of the two vectors u,v,w. The symbol Cu^3 denotes Cuuu and Cu^2v denotes Cuuv, etc. This notation is more convenient for the discussion of this section than the one used in Section 10.

In order to be specific, let us first discuss the bifurcations in Equation (16.1) for (λ,ν) with $\lambda \geqslant 0$ and (λ,ν) near zero. Define a mapping $G: \mathbb{R}^2 \times \mathbb{R} \times \mathbb{R} \to \mathbb{R}^2$ by

$$\begin{aligned} G(u,s,t) &= F(u,\frac{s^2}{2},\frac{t^3}{6}) \\ &= \frac{1}{6}Cu^3 + \frac{s^2}{2}Lu + \frac{t^3}{6}k + \text{h.o.t.} \end{aligned} \tag{16.2}$$

Studying the solutions of the equation

$$G(u,s,t) = 0 \tag{16.3}$$

is equivalent to studying the solutions of Equation (16.1) for $\lambda \geqslant 0, \nu$ near $(0,0)$. The function G satisfies

$$\begin{aligned} &G(0) = 0,\ G'(0)(u,s,t) = 0,\ G''(0)(u,s,t)^2 = 0, \\ &G'''(0)(u,s,t)^3 = Cu^3 + 3s^2Lu + t^3k. \end{aligned} \tag{16.4}$$

Under some reasonable hypothesis, the solutions of Equation (16.3) should be close to the solutions of the equation

$$G'''(0)(u,s,t)^3 = 0. \tag{16.5}$$

It is precisely this hypothesis that we wish to discover.

If (u_0,s_0,t_0) is a solution of Equation (16.5), define the linear map $B: \mathbb{R}^4 \to \mathbb{R}^2$ by the relation

$$\begin{aligned} B(u,s,t) &= G'''(0)(u_0,s_0,t_0)^2(u,s,t) \\ &= Cu_0^2u + s_0Lu + 2s_0sLu_0 + t_0^2\,tk. \end{aligned} \tag{16.6}$$

Since B takes $\mathbb{R}^4$ into $\mathbb{R}^2$, it must have null space of at least dimension two. Therefore, the most natural hypothesis to make is that B satisfies

(A) If (u_0, s_0, t_0) satisfies Equation (16.5) and $B: \mathbb{R}^4 \to \mathbb{R}^2$ is defined by Relation (16.6), then $\dim \mathfrak{R}(B) = 2$.

When Hypothesis (A) is satisfied Magnus says that the solution (u_0, s_0, t_0) is nondegenerate.

If Hypothesis (A) is satisfied, then one can apply the implicit function theorem to solve Equation (16.3) for two of the collection of four variables (u,s,t) in a neighborhood of (u_0, s_0, t_0). However, the variables (s,t) play a special role in the physical problem. We want to be able to determine the curves in (s,t) space across which the number of solutions change. Hypothesis (A) gives no special preference to these variables and therefore is probably not sufficient to determine the bifurcation curves.

If one returns to Section 10, one can show that the following hypotheses are equivalent to (H1) - (H3):

(H1) $Cu^3 = 0$ implies $u = 0$.

(H2') If $\frac{1}{6} Cu^3 \pm Lu = 0$, then $\det (Cu^2 \pm L) \neq 0$.

(H3') If $H = \frac{1}{6} Cu^3 + sLu + 2k = 0$, $\Delta = \det(\partial H/\partial u) \neq 0$, then $\det \partial(H,\Delta)/\partial(u,s) \neq 0$.

One can now show that Hypotheses (H1), (H2'), (H3') imply Hypothesis (A). Further research is needed to determine when one has equivalence between these hypotheses.

17 BIFURCATION FROM FAMILIES OF SOLUTIONS - APPLICATIONS TO NONLINEAR OSCILLATIONS

In previous sections, we have given a procedure for analyzing all of the bifurcations near the zero solution of families of nonlinear equations

$$M(x,\lambda) = 0. \tag{17.1}$$

The complete analysis was possible because of certain generic hypotheses on $M(x,\lambda)$. One of the hypotheses involved the lowest order terms of $M(x,0)$ and implied in particular that $x = 0$ is an isolated solution of

$$M(x,0) = 0. \tag{17.2}$$

In this section, we discuss some recent results of Hale [32] on bifurcation near families of solutions. They generalize those of Hale and Táboas [34] for a problem in nonlinear oscillations in ordinary differential equations which in turn were motivated by Loud [57].

Suppose Equation (17.2) has a one parameter family of solutions $x = p(t)$, $0 \leqslant t \leqslant 1$, where p is continuous with derivatives up through order two and $p'(t) = dp(t)/dt \neq 0$, $0 \leqslant t \leqslant 1$. To avoid a special treatment near $t = 0$ and $t = 1$, we assume $p(0) = p(1)$, $p'(0) = p'(1)$, $p''(0) = p''(1)$. Thus, we may assume p is 1-periodic and $p \in C^2(\mathbb{R}, X)$. If $\Gamma = \{p(t),\ 0 \leqslant t \leqslant 1\} \subseteq X$, the problem is to characterize the solutions (x,λ) of (17.1) in a neighborhood of $\Gamma \times \{0\} \subseteq X \times \Lambda$. Under certain hypotheses, we give such a characterization.

Since $M(p(t),0) = 0$ for $0 \leqslant t \leqslant 1$, it follows that $p'(t)$ is a nonzero element of the null space $\mathfrak{N}(A(t))$ of the linear operator

$$A(t) = \partial M(p(t),0)/\partial x$$

for $0 \leqslant t \leqslant 1$. Suppose that

$$\dim \mathfrak{N}(A(t)) = 1 = \operatorname{codim} \mathcal{R}(A(t)) \tag{17.3}$$

for $0 \leqslant t \leqslant 1$ and $\Lambda = \mathbb{R}^2$. For any element $x \in X$, let $[x]$ denote the span of X. Since $p'(t) \neq 0$, Hypothesis (17.3) implies $[p'(t)] = \mathfrak{N}(A(t))$. Also, suppose $q \in C^2(\mathbb{R}, Z)$ is such that $[q(t)] \oplus \mathcal{R}(A(t)) = Z$. If $B(X)$ denotes the Banach space of bounded linear operators on X, let $U \in C^2(\mathbb{R}, (X))$ be such that $U(t)$ is a projection onto $\mathfrak{N}(A(t))$ and $E \in C^2(\mathbb{R}, B(Z))$, $E(t)$ a projection onto $\mathcal{R}(A(t))$, $I-E(t)$ a projection onto $[q(t)]$. We also suppose U, E are 1-periodic.

If $\Lambda = \mathbb{R}^2$, $\lambda = (\lambda_1, \lambda_2) \in \Lambda$, define $\alpha_j \in C^2(\mathbb{R}, \mathbb{R})$, $j = 1,2$, 1-periodic, by the relation

$$\alpha_j(t)q(t) = (I-E(t))\partial M(p(t),0)/\partial \lambda_j. \tag{17.4}$$

Our first hypothesis on $\alpha(t) = (\alpha_1(t), \alpha_2(t))$ is

(H1) $\alpha(t) \neq 0$ for $t \in \mathbb{R}$.

If $\beta(t) = (\alpha_2(t), -\alpha_1(t))$, then $\beta(t) \neq 0$ by Hypothesis (H1) and we can let $\phi(t)$ be the angle measured in the counterclockwise direction which $\beta(t)$ makes with the horizontal axis. The function $\phi \in C^2(\mathbb{R}, \mathbb{R})$ and is 1-periodic. We impose the following hypotheses on ϕ:

(H2) The function $\phi'(t)$ has at most a finite set of zeros $\{t^k, k = 1,2,\ldots,n\} \subseteq [0,1)$ and $\phi''(t^k) \neq 0$ for $k = 1,2,\ldots,n$.

(H3) $\phi(t^j) \neq (t^k)$, $j \neq k$, $j,k = 1,2,\ldots,n$.

We now state the main results together with implications. Suppose γ is a smooth curve in $\mathbb{R}^2$ through the origin. If for any $q \in \gamma$, $q \neq (0,0)$, $L_q^{\perp}$ denotes a positively oriented normal to L_q at q, we say γ is crossed from right to left at q if γ is crossed by moving along L_q^1 in the positive direction.

Theorem 17.1 If Hypotheses (H1) - (H3) are satisfied, then there exist neighborhoods U of Γ,V of $\lambda = (0,0)$, and $s_0 > 0$, such that, for each $t^j \in \{t^k, k = 1,2,\ldots,n\}$, there corresponds a unique curve $\mathbb{C}_j \subseteq V$, tangent to the line $\alpha(t^j)\cdot\lambda = 0$ at zero, $\mathbb{C}_j \cap \partial V \neq \phi$, each $\mathbb{C}_j$ intersects lines through the origin in at most one nonzero point, these curves intersect only at (0,0), the number of solutions of Equation (17.1) increases (or decreases) by exactly two as $\mathbb{C}_j$ is crossed from right to left if t^j is a relative minimum (or maximum) of ϕ.

The curves $\mathbb{C}_j$ can be defined parametrically in the form $\lambda = s\beta_j(s)$, $0 \leqslant s < s_0$ where $\beta_j \in C^2([0,s_0),\mathbb{R}^2)$, $|\beta_j(s)| = 1$, $0 \leqslant s < s_0$, and $\alpha(t^j)\cdot\beta_j(0) = 0$. If t_*, t^* are the absolute minimum, maximum, respectively, of ϕ and

$$S(V) = \{\lambda \in V\colon \lambda\cdot\beta^*(s) < 0 < \lambda\cdot\beta_*(s),\ 0 \leqslant s < s_0\},$$

where $\lambda = s\beta^*(s)$, $\lambda = s\beta_*(s)$ are the curves corresponding to t^* and t_*, then there are no solutions of Equation (17.1) in U for $\lambda \in S(V)$, at least two in $S^c(V) = V\setminus S(V)$ and all solutions are distinct in the interior of $S^c(V)$.

The curves $\mathbb{C}_j$ in Theorem 17.1 are the bifurcation curves. To see how easy it is to obtain the complete qualitative picture of the bifurcations near $\lambda = 0$, let us consider a few special cases. If $\phi(t)$ has only one maximum at t^* and one minimum at t_*, there are only two bifurcation curves $\mathbb{C}_*$, $\mathbb{C}^*$ corresponding to t_*, t^*, respectively. There are no solutions of Equation (17.1) in U for $\lambda \in S(V)$ and exactly two solutions in $S^c(V)$ which are distinct in the interior of $S^c(V)$ (see Figure 16). If there are two maxima and two minima (ther must always be an even number of maxima and minima by periodicity), then the situation is depicted in Figure 17. By changing the function

ϕ, one can obtain every possible rotation of these pictures.

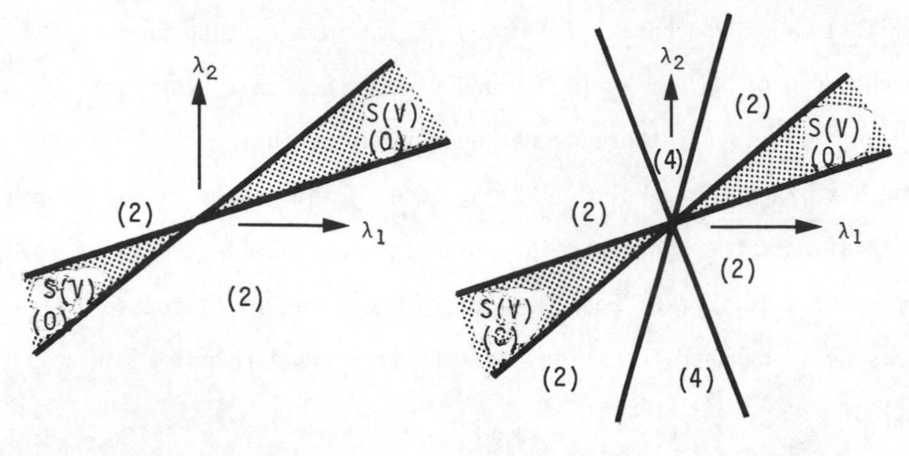

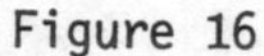
Figure 16

Figure 17

Another interesting special case is $\alpha_1(t) = -1$, $t \in \mathbb{R}$. Hypothesis (H1) is always satisfied, $\beta(t) = (\alpha_2(t), 1)$, $\phi(t) = \cot^{-1}\alpha_2(t)$ and the hypotheses (H1), (H2) are equivalent to

(H2') The function α_2' has a most finite set of zeros $\{t^k, k = 1,2,\ldots,n\} \subseteq [0,1)$ and $\alpha_2''(t^k) \neq 0, k = 1,2,\ldots,n$.

(H3') $\alpha_2(t^j) \neq \alpha_2(t^k)$, $j \neq k$, $j, k = 1,2,\ldots,n$.

Theorem 17.1 for this case is essentially contained in [34]. Since $\alpha_1(t) = -1$, the set S(V) must contain the λ_1-axis and, thus, the bifurcation diagram is a rotated version of the ones in Figures 16 and 17.

The following result gives some information about the possible behavior of the solutions of Equation (17.1) as $\lambda \to 0$.

Theorem 17.2 Suppose Hypotheses (H1) - (H3) are satisfied, U,V are the neighborhoods given in Theorem 17.1, and suppose γ is a continuous curve defined parametrically by $\lambda_1 = \lambda_1(\tau)$, $\lambda_2 = \lambda_2(\tau)$, $0 \leq \tau \leq 1$, and $\lambda_1^2(\tau) + \lambda_2^2(\tau) = 0$ if and only if $\tau = 0$. Also, suppose $\gamma \subseteq V$ and for each point $(\lambda_1(\tau), \lambda_2(\lambda) \in \gamma$, there is a solution $x(\tau) \in U$ of Equation (17.1) which is continuous in τ on the half open interval (0,1]. If

$$\mathscr{S}(\gamma) = \{x(\tau),\ 0 < \tau \leq 1\} \subseteq X \tag{17.5}$$

is precompact, and

$$\begin{aligned} \phi_m(\gamma) &= \liminf_{\tau\to 0} \cot^{-1}(\lambda_1(\tau)/\lambda_2(\tau)), \\ \phi_M(\gamma) &= \limsup_{\tau\to 0} \cot^{-1}(\lambda_1(\tau)/\lambda_2(\tau)), \end{aligned} \tag{17.6}$$

then there is an interval $I(\gamma) \subseteq [0,1]$ such that $\phi(I(\gamma)) = [\phi_m(\gamma), \phi_M(\gamma)]$ and

$$(C\ell\, \mathscr{S}(\gamma)) \setminus \mathscr{S}(\gamma) = \{p(t),\ t \in I(\gamma)\}. \tag{17.7}$$

A consequence of the above result is the following:

Corollary 17.1 If $\gamma, x(\tau)$ satisfy the conditions of Theorem 17.2, then a necessary and sufficient condition that $x(\tau)$ have a limit as $\tau \to 0$ is that $\cot^{-1}(\lambda_1(\tau)/\lambda_2(\tau))$ has a limit ϕ_0 as $\tau \to 0$. In this case, $x(\tau) \to p(t_0)$ where $t_0 \in [0,1)$ is a solution of the equation $\cot^{-1}(\lambda_1(t)/\lambda_2(t)) = \phi_0$.

The fact that one can obtain solutions which are not continuous in λ at $\lambda = 0$ is not surprising. Consider the scalar equation $\lambda_1 x - \lambda_2 = 0$ which has the solution $x = \lambda_2/\lambda_1$ for $\lambda_1 \neq 0$. Along a curve $\gamma \in \mathbb{R}^2$, this solution has a limit as $\lambda \to 0$ in γ if and only if λ_1/λ_2 approaches a limit as $\lambda \to 0$ in γ.

Let us now make a more interesting application to the second order scalar ordinary differential equation

$$\frac{d^2x}{ds^2} + g(x) + \lambda_1 h(s)\frac{dx}{ds} - \lambda_2 f(s) = 0 \tag{17.8}$$

where h,f are continuous and 1-periodic, $g \in C^2(\mathbb{R},\mathbb{R})$, $xg(x) > 0$ for $x \neq 0$. For $\lambda_1 = \lambda_2 = 0$, the equation

$$\frac{d^2x}{ds^2} + g(x) = 0 \tag{17.9}$$

has a general solution of the form $x = \psi(\omega(a)s + t, a)$, $(a,t) \in \mathbb{R}^2$, where $\psi(\zeta,a) = \psi(\zeta+1,a)$ for all (ζ,a), and $(a,0) = (x(0), dx(0)/ds)$. We suppose Equation (17.9) has a nondegenerate 1-periodic orbit; that is,

$$\text{There is an } a_0 > 0 \text{ such that } \omega(a_0) = 1,\ d\omega(a_0)/da \neq 0. \tag{17.10}$$

Let $Z = \{y: \mathbb{R} \to \mathbb{R}$ which are continuous and 1-periodic$\}$ and use the supremum norm on Z. Let $X = \{y \in Z$: y has continuous derivatives up through order two$\}$ and use the usual C^2 norm on X. If we define $M: X \times \Lambda \to Z$, $\Lambda = \mathbb{R}^2$ by

$$M(x,\lambda)(s) = \frac{d^2x(s)}{ds^2} + g(x(s)) + \lambda_1 h(s)\frac{dx(s)}{ds} - \lambda_2 f(s)$$

then we are in a position to apply the previous results. In fact, if $p(t)(s) = \psi(\omega(a_0)s + t, a_0)$, then $p(t) \in X$ and satisfies $M(p(t),0) = 0$, $0 \leq t \leq 1$. Also, Hypothesis (17.10) implies that $\dim \mathfrak{N}(A(t)) = 1 = \operatorname{codim} \mathcal{R}(A(t))$, where $A(t) = \partial M(p(t),0)/\partial x$. Furthermore, the function $\dot{p}(t)$ is a basis for $\mathfrak{N}(A(t))$ and a complement for $\mathcal{R}(A(t))$. It is now an obvious calculation to see that the functions $\alpha_1(t)$, $\alpha_2(t)$ in (17.4) are given by

$$\alpha_1(t) = -\int_0^1 h(s)\dot{p}(s+t)^2 ds,\quad \alpha_2(t) = \int_0^1 \dot{p}(s+t) f(s) ds. \tag{17.11}$$

If (α_1,α_2) satisfy (H1) - (H3), then the above results are directly applicable to the determination of the bifurcation curves for the 1-periodic solutions of Equation (17.8) which lie in a neighborhood of the periodic orbit

$\Gamma \subseteq \mathbb{R}^2$ of Equation (17.9) defined by $\Gamma = \{(p(s), dp(s)/ds), 0 \leq s \leq 1$. For $h(s) = 1$, $0 \leq s \leq 1$, these results were previously obtained in [34].

Theorem 17.2 states the general situation for the parameters (λ, μ) varying independently as they approach zero, the solutions of Equation (17.8) will not remain close to any particularl $p(\cdot + \alpha_0)$ but to the family of periodic functions $\{p(\cdot + \alpha), \alpha \in I_\gamma$ (see Figure 18).

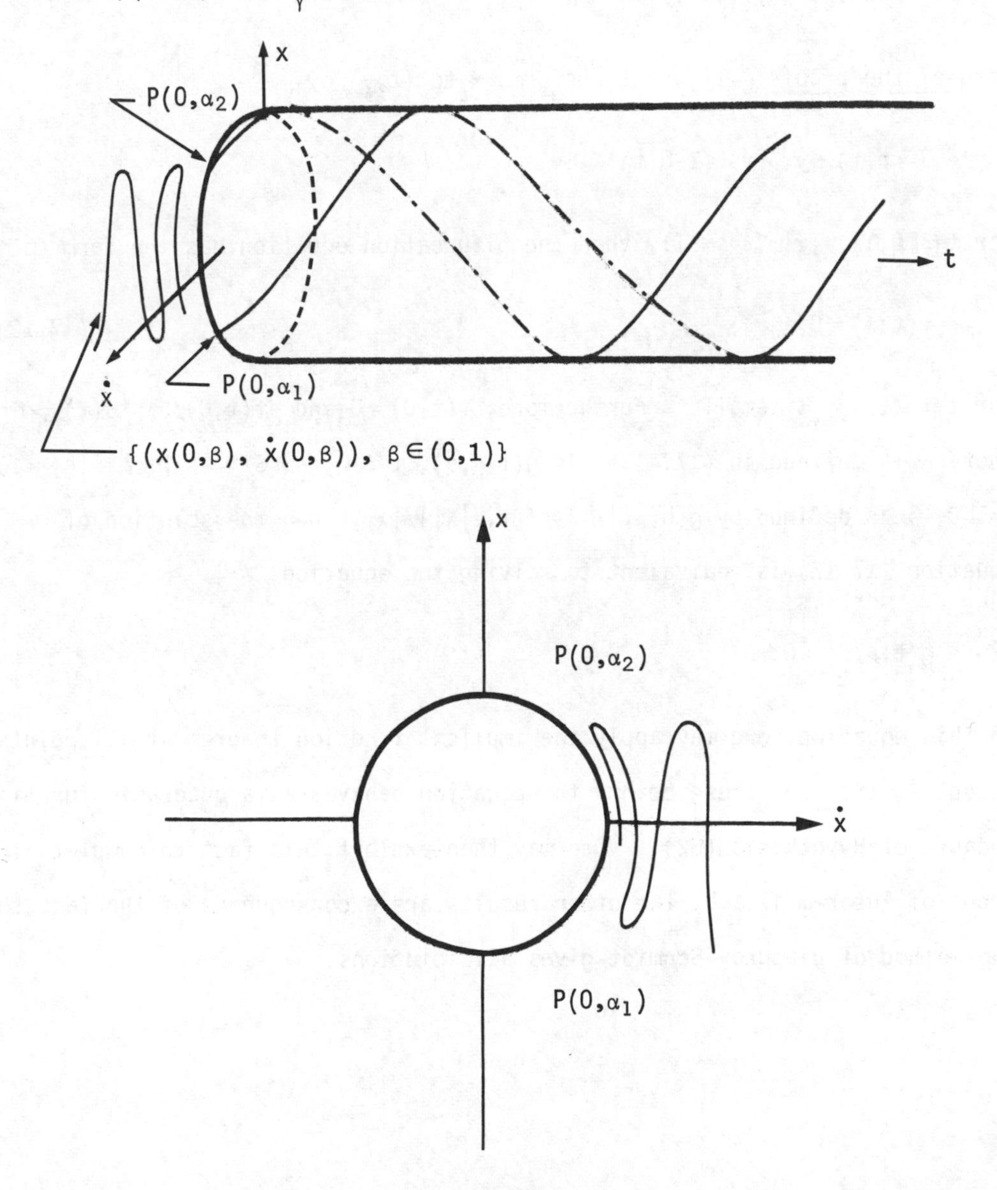

Figure 18

Since one can obviously choose a curve such that $I_\gamma = [0, 1]$, one has a curve of 2π-periodic solutions of Equation (17.8) with the property that the corresponding initial values converge to the entire cylinder in Figure 18. If one remembers the fact that damping in nonlinear oscillations generally induces a phase shift, then the results do not seem as surprising and probably can be observed physically.

Idea of the proofs: If one transforms x to (t,y) by

$$X = p(t) + y, \quad y \in (I-U(t))X,$$

for $t \in [0,1)$, $|y| < \delta$ small, then the bifurcation equation has the form

$$f(t,\lambda) = 0, \tag{17.12}$$

for $t \in \mathbb{R}$, $\lambda \in \mathbb{R}^2$ small. Furthermore, $f(t,0) = 0$ and $\partial f(t,0)/\partial\lambda = \alpha(t) \in \mathbb{R}^2$ where α is defined in (17.4). If $g(t,\beta,s)$, $t \in \mathbb{R}$, $\beta \in S^1 = \{\beta \in \mathbb{R}^2 : |\beta| = 1\}$, $s \in [0,\infty)$ is defined by $g(b,\beta,|\lambda|) = f(t,\beta|\lambda|)/|\lambda|$, then the solution of Equation (17.12) is equivalent to solving the equation

$$g(t,\beta,s) = 0.$$

To this equation, one may apply the implicit function theorem at all points except $t = t^j$. At these points the equation behaves as a quadratic function because of Hypothesis (H2). One may then exploit this fact to complete the proof of Theorem 17.1. The other results are a consequence of the fact that the method of Liapunov-Schmidt gives all solutions.

18 HARMONICS IN THE DUFFING EQUATION

The purpose of this section is to discuss another type of bifurcation problem for the equation $M(x,\lambda) = 0$ when there is a family of solutions of the equation $M(x,0) = 0$. More specifically, if $M(x,\lambda) = Ax - \lambda N(x)$ where A is a linear operator $\dim \mathfrak{N}(A) \geq 2$, then the equation $M(x,0) = 0$ has atleast a two dimensional subspace of solutions. This problem is distinctively different from the one of the previous section. Probably the most important distinction is that the family of solutions of the equation $M(x,0) = 0$ is noncompact. This means that one must allow for the possibility that some solutions of $M(x,\lambda) = 0$ may approach infinity as $\lambda \to 0$. The generic hypotheses on the equation must be imposed in such a way as to permit one to obtain a priori bounds on the manner in which solutions can become unbounded. We do not understand this problem very well and discuss only an example; namely the 2π-periodic solutions of the Duffing equation

$$\frac{d^2u}{dt^2} + u = p_1 u + p_4 \frac{du}{dt} - p_2 u^3 + p_3 \cos t, \qquad (18.1)$$

where $p = (p_1, p_2, p_3, p_4)$ is a real four dimensional vector varying in a neighborhood U of the origin. The results given below are due to Hale and Rodrigues [33]. The term $p_3 \cos t$ could be replaced by $p_3 f(t)$ where

$$\left(\int_0^{2\pi} f(t) \sin t \, dt\right)^2 + \left(\int_0^{2\pi} f(t) \cos t \, dt\right)^2 \neq 0,$$

and the conclusions would be the same. The term cost is chosen only for simplicity.

Along with Equation (18.1), we consider

$$\frac{d^2u}{dt^2} + u - p_1u + p_2u^3 = 0. \tag{18.2}$$

A <u>2π-periodic solution continuous</u> in p_3, p_4 is a continuous function from a deleted neighborhood $V - \{(0,0)\}$ (depending on p_1,p_2) of $0 \in \mathbb{R}^2$ into the space of 2π-periodic functions which associates to each $(p_1,p_2) \in V - \{(0,0)\}$ a 2πperiodic solution $u(p_1,p_2,p_3,p_4)(t)$ of Equation (18.1). Furthermore, the set $\{u(p_1,p_2,p_3,p_4),\ (p_3,p_4) \in V - \{(0,0)\}\}$, with the uniform topology, is precompact and every limit point of this set as $(p_3,p_4) \to 0$ is either a constant function or a periodic solution of (18.2) of least period 2π.

The idea for considering this particular type of continuous dependence on the parameters came from a paper of Hale and Táboas [34] in the consideration of 2π-periodic solutions of another type of equation.

To obtain a priori bounds on the solutions of Equation (18.2), we restrict the discussion to 2π-periodic solutions continuous in (p_3,p_4). It is therefore necessary to discuss some detailed properties of the solutions of Equation (18.2). The necessary information is contained in the following lemma.

<u>Lemma 18.1</u> There is a constant $k > 0$ and a neighborhood U in $\mathbb{R}^2$ of $(p_1,p_2) = (0,0)$ such that a nonconstant 2π-periodic solution u of (18.2) exists for $(p_1,p_2) \in U$ if and only if $p_1p_2 > 0$, this solution is unique except for a phase shift and satisfies

$$|u(t)| \leq k|p_1/p_2|^{1/2}.$$

If $(p_1,p_2) \in U$ and either $p_1p_2 \leq 0$, $p_2 > 0$ or $p_1 \neq 0$, $p_2 = 0$, the only 2π-periodic solution of (18.2) is $u = 0$. If $p_1p_2 \neq 0$, $p_2 < 0$, there are the additional 2π-periodic solutions $u = 0$, $u = \pm[1-p_1)/|p_2|]^{1/2}$. If $p_1 = 0$, $p_2 = 0$, every solution of (18.2) is 2π-periodic.

Ideas of the proof: Every periodic orbit of Equation (18.2) encircles the origin in the $(u, du/dt)$ plane. If $(r,0)$ designates the point at which the orbit intersects the u-axis, then the period $\tau(r,p_1,p_2)$ is a strictly monotone function of r and there is an $r_0 > 0$ such that $\tau(r_0,p_1,p_2) = 2\pi$ if and only if $p_1p_2 > 0$. If $p_1p_2 > 0$, then $r_0 = r_0(p_1,p_2)$ is unique. If the period $\tau(r,p_1,p_2)$ is represented in terms of elliptic integrals, then one observes that $r_0(p_1,p_2) \to \infty$ as $p_2 \to 0$. One now uses the scaling $u = v(p_1/p_2)^{1/2}$ for $p_1p_2 > 0$ to obtain

$$\ddot{v} + v = p_1(v - v^3).$$

Standard arguments in classical perturbation theory (see [30]) yield the estimates of the lemma. The other statements are easy to verify to complete the proof.

For $p_2 < 0$, there are always two 2π-periodic solutions of Equation (18.1) which exist for $p = (p_1,p_2,p_3,p_4) \in \mathbb{R}^4$ in a neighborhood of zero and coincide with $\pm[(1-p_1)/|p_2]^{1/2}$ for $p_3 = p_4 = 0$. This is proved by the implicit theorem and is stated precisely in the following lemma.

Lemma 18.2 There is a neighborhood U in $\mathbb{R}^4$ of $p = 0$ and that for every $p \in U$, $p_2 < 0$, there exist two 2π-periodic solutions of Equation (18.1) which are continuous in (p_3,p_4) and coincide with $\pm[(1-p_1)/|p_2|]^{1/2}$ for $p_3 = p_4 = 0$. All other such 2π-periodic solutions of Equation (18.1) for $p_3 = p_4 = 0$ must coincide with the solution $u = 0$ of Equation (18.2) or a nonconstant periodic solution of Equation (18.2).

To obtain the other 2π-periodic solutions of Equation (18.1) continuous in (p_3,p_4), let $p_2 > 0$, $u = vp_2^{-1/2}$ in Equation (18.1) to obtain the equation

$$\frac{d^2v}{dt^2} + v = p_1 v + p_4 \frac{dv}{dt} - v^3 + \sigma \cos t, \tag{18.3}$$

where $\sigma = p_2^{1/2} p_3$. From Lemma 18.1, we know that $p_2^{1/2} u$ is bounded and, furthermore, that the only 2π-periodic solutions of Equation (18.3) that need to be considered are those for which v is small. If $p_2 < 0$, the same remark is true (of course, with $-v^3$ replaced by $+v^3$) provided the two solutions in Lemma 18.2 are excluded from the discussion.

We now discuss 2π-periodic solutions of Equation (18.3) for (v, p_1, σ, p_4) in a small neighborhood of the origin. We are going to apply the method of Liapunov-Schmidt, but include the details since a few modifications are required. Any 2π-periodic solution of Equation (18.3) for $p_1 = p_4 = \sigma = 0$ must be equal to $r\cos(t-\phi) + O(r^2)$ for some constants r, ϕ, $-\pi/2 \leqslant \phi \leqslant \pi/2$. Therefore, by letting $t \to t + \phi$, we will obtain a solution of our problem by considering the 2π-periodic solutions of the equation

$$\frac{d^2v}{dt^2} + v = p_1 v + p_4 \frac{dv}{dt} - v^3 + \sigma \cos(t + \phi), \tag{18.4}$$

which, for $p_1 = p_4 = \sigma = 0$, are equal to $r\cos t + O(r^2)$.

Let $\mathscr{S} = \{h: \mathbb{R} \to \mathbb{R}:$ h is continuous $h(t + 2\pi) = h(t)\}$ and for any $h \in \mathscr{S}$, let $|h| = \sup_t |h(t)|$. Let $P: \mathscr{S} \to \mathscr{S}$ be the projection defined by

$$(Ph)(t) = \frac{1}{\pi} \cos t \int_0^{2\pi} h(s) \cos s \, ds + \frac{1}{\pi} \sin t \int_0^{2\pi} h(s) \sin s \, ds. \tag{18.5}$$

For any $h \in \mathscr{S}$, the equation

$$\frac{d^2v}{dt^2} + v = h \tag{18.6}$$

has a solution in $\mathscr{S}$ if and only if $Ph = 0$. Furthermore, there is a continuous linear operator $K: (I-P)\mathscr{S} \to (I-P)\mathscr{S}$ such that $K(I-P)h$ is the unique solution of

$$\frac{d^2v}{dt^2} + v = (I-P)h \tag{18.7}$$

which satisfies $PK(I-P)h = 0$; that is, $K(I-P)h$ is simply the 2π-periodic solution of Equation (18.6) which does not contain cost, sint in its Fourier series. To this solution $K(I-P)h$, one can add an arbitrary linear combination of sint and cost to obtain the general solution of Equation (18.7). As remarked earlier, it is only necessary for us to add a term $r\cos t$.

If r is fixed and we define

$$P_r = \{h \in P: (Ph)t = r\cos t\},$$
$$f(v,p_1,p_4) = p_1 v + p_4 \frac{dv}{dt} - v^3,$$

then v is a solution of Equation (18.4) in P_r if and only if

$$v = r\cos t + w, \quad w \in (I-P)\mathscr{S}, \tag{18.8a}$$

$$\frac{d^2w}{dt^2} + w = (I-P)f(r\cos t(\cdot) + w, p_1, p_4), \tag{18.8b}$$

$$P[f(r\cos(\cdot) + w, p_1, p_4) + \sigma\cos(\cdot + \phi)] = 0, \tag{18.8c}$$

since $(I-P)\cos(\cdot + \phi) = 0$.

By an application of the implicit function theorem, there are $\delta > 0$, $\varepsilon > 0$, such that, for $|r| < \delta$, $|p_1| + |p_4| < \varepsilon$, there is a unique solution $w^*(r,p_1,p_4)$ of Equation (18.8b) in $(I-P)\mathscr{S}$, the function $w^*(r,p_1,p_4)$ is analytic in $r,p_1,p_4)$ and $w^*(0,p_1,p_4) = 0$. Furthermore, it is very easy to see that $w^*(r,p_1,0)$ is an even function of t since $v = r\cos t + w$ and only powers of v occur on the right hand side of Equation (18.8b) for $p_4 = 0$.

Since $w^*(r,p_1,p_4)$ is uniquely determined it follows that there is a solution $v = r\cos t + w \in P_r$ of Equation (18.4) which lies in a sufficiently small neighborhood of zero if and only if $v = r\cos t + w^*(r,p_1,p_4)$ and (r,ϕ,p_1,σ,p_4) satisfy the bifurcation equations

$$P[f(r\cos(\cdot)+w^*(r,p_1,p_4),p_1,p_4)+\sigma\cos(\cdot+\phi)]=0.$$

From the definition of P in Equation (18.5), these latter equations are equivalent to the system of equations

$$G_1(r,\phi,p_1,\sigma,p_4) \overset{\text{def}}{=} \sigma\cos\phi+\frac{1}{\pi}\int_0^{2\pi} f(r\cos t+w^*(r,p_1,p_4)(t),p_1,p_4)\cos t\,dt=0,$$

$$G_2(r,\phi,p_1,\sigma,p_4) \overset{\text{def}}{=} \sigma\sin\phi+\frac{1}{\pi}\int_0^{2\pi} f(r\cos t+w^*(r,p_1,p_4)(t),p_1,p_4)\sin t\,dt=0.$$

Since $w^*(r,p_1,0)$ is an even function, it is clear that $G_2(r,\phi,p_1,\sigma,0)=\sigma\sin\phi$. For $w^*(r,p_1,p_4)=0$, it is easy to evaluate the above integrals. If this computation is made and one uses the Taylor series to obtain the order estimate $O(|p_1r|+|p_4r|+|r|^3)$ on $w^*(r,p_1,p_4)$, the bifurcation equations become

$$G_1(r,\phi,p_1,\sigma,p_4)=\sigma\cos\phi+p_1r-\frac{3}{4}r^3+rg_1(r,p_1,p_4)=0, \tag{18.9a}$$

$$G_2(r,\phi,p_1,\sigma,p_4)=\sigma\sin\phi+p_4r+p_4rg_2(r,p_1,p_4)=0, \tag{18.9b}$$

where

$$g_1(r,p_1,p_4)=O(|p_1|^2+|p_4|^2+|p_1p_4|+r^2|p_1|+r^2|p_4|+r^4), \tag{18.10}$$
$$g_2(r,p_1,p_4)=O(r^2+|p_1|+|p_4|).$$

These results are summarized in the following lemma.

Lemma 18.3 There is a neighborhood $U\subset\mathbb{R}^4$ of $(r,p_1,\sigma,p_4)=0$ and a neighborhood $V\subset\mathscr{S}$ of $v=0$ such that Equation (18.4) has a solution $v\in P_r\cap V$ for $(r,p_1,\sigma,p_4)\in U$ and a given ϕ if and only if (r,p_1,σ,p_4,ϕ) satisfy the bifurcation Equations (18.9) where g_1, g_2 satisfy (18.10).

An immediate corollary is the following result on the undamped Duffing equation.

Corollary 18.1 There is a neighborhood $U \subset \mathbb{R}^2$ of $(p_1,\sigma) = 0$ and a neighborhood $V \subset \mathscr{S}$ of $v = 0$ such that the only 2π-periodic $\mathscr{S}$ solutions in V of the undamped Duffing equation

$$\frac{d^2v}{dt^2} + v = p_1 v - v^3 + \sigma \cos t \tag{18.11}$$

are even functions of t if $\sigma \neq 0$.

Proof. For $p_4 = 0$, the second bifurcation Equation (18.9b) is $\sigma \sin\phi = 0$. Therefore, if $\sigma \neq 0$, we must have $\phi = 0$. If $\phi = 0$, then $v(t) = r\cos t + w^*(r,p_1,0)(t)$ is an even solution of Equation (18.11). Since all solutions of Equation (18.11) can be obtained from Equation (18.4) and the Lemma 18.3, we have proved the desired result.

Remark 18.1 The conclusion of Corollary 18.1 is also valid for the equation

$$\frac{d^2v}{dt^2} + v = p_1 v - v^3 + \sigma f(t)$$

where f(t) is even in t and $\int_0^{2\pi} f(t)\cos t\,dt = \pi$. To prove this, one considers the same equation with t replaced by $t + \phi$ and observes that the bifurcation equation obtained by projecting onto sint has the form $\sigma(\sin\phi)h(r,\phi,p_1) = 0$ with $h(0,\phi,0) = 1$. One can now repeat the same argument as in the proof of Corollary 18.1 to complete the proof.

The ideas used in the proof of the evenness of the solutions in Corollary 18.1 has been abstracted to obtain interesting information about the solutions of equations in Banach spaces which remain invariant under certain groups of transformations (see Rodrigues and Vanderbauwhede [70]). This latter paper also contains interesting applications to ordinary and partial differential equations.

It remains to analyze the bifurcation equations (18.9). This is not an easy task and one must exploit the explicit form given in Equations (18.9) and obtained from the inherent symmetry in the unforced Duffing equation. The results are now summarized and the reader is referred to [33] for the detailed analysis.

Theorem 18.1 Let $\sigma = p_2^{1/2} p_3$. There is a neighborhood U in $\mathbb{R}^3$ of (0,0,0) such that the bifurcation surface Γ for Equation (18.1) with $(p_1, \sigma, p_4) \in U$ is depicted in Figure 19 and the surface for $p_2 > 0$ is approximately given by the equation

$$\sigma^2 = \frac{8}{81}[p_1^3 + 9 p_1 p_4^2 \pm (p_1^2 - 3p_4^2)^{3/2}].$$

The number of 2π-periodic solutions of Equation (18.1) at a point $(p_1, \sigma, p_4) \in U$ which are continuous in (p_3, p_4) is shown in Figure 19.

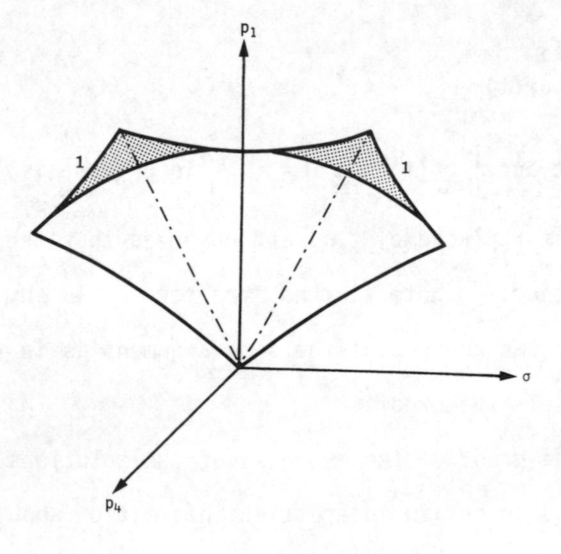

Figure 19

For other discussions of nonlinear oscillations and generic bifurcation, see Holmes and Rand [35] and Takens [82]. Other interesting problems in Hazashi [29] possibly could be understood better using these methods.

References

1 J. A. Alexander and J. Yorke, Global bifurcation of periodic orbits. U. of Maryland report (1974).

2 Antonio Ambosetti, Some remarks on the buckling problem for a thin clamped shell. Richerche di Matematica 23, 161-170.

3 A. Ambrosetti and G. Prodi, On the inversion of some differentiable mappings with singularities between Banach spaces. Annali Mat. Pura Appl. 93 (1972) 231-246.

4 V. I. Arnold, Lectures on bifurcation in versal families. Russian Math. Surveys, 27 (1972) 54-123.

5 S. Bancroft, J. K. Hale and D. Sweet, Alternative problems for nonlinear functional equations. J. Diff. Eqns. 4 (1968) 40-56.

6 L. Bauer, E. Reiss and H. Keller, Multiple eigenvalues lead to secondary bifurcations. SIAM Review 17 (1975) 101-122.

7 L. Bauer and E. L. Reiss, Numerical bifurcation and secondary bifurcation - case history, p. 443-467 of Numerical Solutions of Partial Differential Equations, III. (1976) Academic Press.

8 M. Berger and P. C. Fife, On von Kármán's equations and the buckling of a thin elastic plate, I, The clamped plate. Comm. Pure Appl. Math. 20 (1967), 687-719.

9 M. S. Berger, Applications of global analysis to specific nonlinear eigenvalue problems. Rocky Mountain J. Math. 3 (1973) 319-354.

10 M. Berger and P. C. Fife, On von Kármán's equations and the buckling of a thin elastic plate, II, Plate with general edge conditions. Comm. Pure Appl. Math. 21 (1968) 227-241.

11 L. Cesari, Functional analysis, nonlinear differential equations and the alternative method. pp 1-198 in Nonlinear Functional Analysis and Differential Equations (1976) Dekker, New York.

12 L. Cesari, Functional analysis and Galerkin's method. Mich. Math. J., 11 (1964) 385-418.

13 N. Chafee, Generalized hopf bifurcation and perturbation in a full neighborhood of a given vector field. (to appear).

14 D. Chillingworth, The catastrophe of a buckling beam. In Dynamical Systems - Warwick 1974, 88-91. Lecture Notes in Math. Vol. 488, Springer (1975).

15 S. Chow and J. Mallet-Paret, Fuller index and global Hopf bifurcation. J. Differential Eqns.(to appear).

16 S. Chow, J. K. Hale and J. Mallet-Paret, Applications of generic bifurcation. I, Arch. Rational Mech. Anal. 59 (1975) 159-188.

17 S. Chow, J. K. Hale and J. Mallet-Paret, Applications of generic bifurcation. II, Arch. Rational Mech. Anal. 62 (1976) 209-235.

18 D. S. Cohen, Multiple solutions of nonlinear partial differential equations. Springer-Verlag Lecture Notes in Mathematics (1973) 322, 15-77.

19 M. G. Crandall and P. H. Rabinowitz, Bifurcation from simple eigenvalues. J. Func. Anal. 8 (1971) 321-340.

20 E. Dancer, Bifurcation theory in real Banach space. Proc. London Math. Soc. 23 (1971) 699-734.

21 E. Dancer, Bifurcation theory for analytic operators. Ibid 26 (1973) 359-384.

22 E. Dancer, Global structure of the solutions of nonlinear real analytic eigenvalue problems. Ibid 27 (1973) 747-765.

23 Dynamical Systems, Warwick 1974. Lecture Notes in Math. Vol. 468 Springer.

24 S. Fucik, J. Necas, J. Soucek and V. Soucek, Spectral Analysis of Nonlinear Operators. Lecture Notes in Math. (1973) Vol. 346, Springer-Verlag.

25 R. E. Gaines and J. Mawhin, Coincidence Degree and Nonlinear Differential Equations. Lecture Notes in Math. Vol. 568, Springer.

26 M. Golubitsky and V. Guillemin, Stable Mapping and their Singularities. Springer-Verlag, New York (1973).

27 W. M. Greenlee, Remarks on branching from multiple eigenvalues. Lecture notes in Math. Vol. 322 (1973) 101-12, Springer-Verlag, New York.

28 D. Gromoll and W. Meyer, On differentiable functions with isolated critical points. Topology 8 (1969), 361-369.

29 C. Hayashi, Nonlinear Oscillations in Physical Systems. McGraw-Hill (1964).

30 J. K. Hale, Ordinary Differential Equations, Wiley-Interscience, (1969).

31 J. K. Hale, Applications of Alternative Problems. CDS Lecture Notes, (1971) Div. Appl. Math. Brown Univ., Providence, R.I.

32 J. K. Hale, Bifurcation near families of solutions. Proc. International Conf. on Diff. Eqns. Uppsala (1977) Lecture Notes in Math. Springer.

33 J. K. Hale and H. M. Rodrigues, Bifurcation in the Duffing equation with several parameters, I, II, Proc. Roy. Soc. Edin. Ser. A. (to appear).

34 J. K. Hale and P. Z. Táboas, Interaction of damping and forcing in a second order equation. INA-TMA (to appear).

35 P. J. Holmes and D. A. Rand, Bifurcations of the forced van der Pol oscillator. Preprint.

36 E. Hopf, Abzweigung einer periodischen Losung von einer stationaren Losung eines differential systems. Ber Math-Phys. Sachsische akademie de Wissenschaften Leipzig 94 (1942), 1-22.

37 J. Ize, Bifurcacion global de orbitas periodicas (to appear).

38 J. A. Ize, Bifurcation theory for Fredholm operators, Ph.D. Thesis (1974) NYU.

39 J. P. Keener, Some modified bifurcation problems with applications to imperfection sensitivity in buckling. Ph.D. Thesis (1972) Cal. Inst. Tech., California.

40 J. Keener and H. Keller, Perturbed bifurcation theory. Arch. Rat. Mech. Anal. 50 (1973) 159-175.

41 J. P. Keener, Secondary bifurcation and perturbed bifurcation theory, preprint.

42 J. P. Keener, Perturbed bifurcation theory at multiple eigenvalues. Arch. Rat. Mech. Anal 56 (1974) 348-366.

43 J. B. Keller and S. Antman, Bifurcation Theory and Nonlinear Eigenvalue Problems. Benjamin (1969).

44 K. Kirchgassner, Multiple eigenvalue bifurcation for holomorphic mappings (1971) 69-99. Contributions to Nonlinear Functional Analysis. Ed. E. H. Zarantello, Academic Press.

45 G. H. Knightly and D. Sather, Nonlinear buckled states of rectangular plates, Arch. Rational Mech. Anal. 54 (1974) 356-372.

46 G. H. Knightly and D. Sather, On nonuniqueness of solutions of the von Kármán equations. Arch. Rat. Mech. Anal. 36 (1970) 65-78.

47 G. H. Knightly, Some mathematical problems from plate and shell theory (1976) 245-268. Nonlinear Functional Analysis and Differential Equations, Dekker, New York.

48 W. T. Koiter, Over de stabiliteit van het elastich evenwicht. Thesis Delft. H. J. Paris, Amsterdam (1945), Eng. Transl. Tech. Rpt. AFFDL-TR-70-25, Feb. 1970.

49 W. T. Koiter, Elastic stability and post-buckling behavior (1963) 257-275. Nonlinear Problems, Ed. R. Langer, Univ. Wisc. Press.

50 N. Kopell and L. N. Howard, Bifurcations under nongeneric situations. Advances in Math. 13 (1974) 274-283.

51 V. M. Krasnoselskii, Small solutions of a class of nonlinear operators. Sov. Math. Dokl. 9 (1968) 579-581.

52 V. M. Krasnoselskii, Investigation of the bifurcation of small eigenfunctions in the case of multiple dimensional degeneration. Ibid 11 (1970), 1609-1613.

53 V. M. Krasnoselskii, The projection method for the study of bifurcation of the zero solution of a nonlinear operator equation with multidimensional degeneracy. Ibid 12 (1971) 967-971.

54 M. A. Krasnoselskii, Topological Methods in the Theory of Nonlinear Integral Equations. Pergamon Press (1964).

55 N. H. Kuiper, C^r-functions near non-degenerate critical points. Mimeographed, Warwick University, Coventry (1966).

56 S. List, Generic bifurcation with application to the von Kármán equations, Ph.D. Thesis (1976) Brown Univ.

57 W. S. Loud, Branching of solutions of two parameter boundary value problems for second order differential equations. ICNO, Berlin (1975) (to appear).

58 J. MacBain, Global bifurcation theorems for noncompact operators. Bull. Am. Math. Soc. 80 (1974) 1005-1009.

59 J. B. MacLeod and D. H. Sattinger, Loss of stability at a double eigenvalue. J. Funct. Anal. 14 (1973) 62-84.

60 R. J. Magnus, Topological stability of the solution set of a nonlinear equation. Batelle, Geneva, Math. Report No. 85 (1974).

61 R. J. Magnus, On the local structure of the zero set of a Banach space valued mapping. J. Functional Anal. 22 (1976) 58-72.

62 R. J. Magnus, The reduction of a vector valued function near a critical point. Battelle, Geneva, Math. Report No. 93 (1975).

63 J. Mallet-Paret, Buckling of cylindrical plates with small curvature. Quart. Appl. Math. (to appear).

64 J. E. Marsden and M. F. MacCracken, The Hopf Bifurcation Theorem and its Applications. Appl. Math. Sci. Vol. 19, Springer-Verlag.

65 B. Matkowsky and L. Putnik, Multiple buckled states of rectangular plates. Int. J. Nonlinear Mech. 9 (1974) 89-103.

66 L. Nirenberg Topics in Nonlinear Functional Analysis. Lecture Notes 1973-74 Courant Institute, New York Univ. (1974).

67 R. Nussbaum, A global Hopf bifurcation theorem for retarded functional differential equations. Trans. Am. Math. Soc. (to appear).

68 G. H. Pimbley, Eigenfunction Branches of Nonlinear Operators and their bifurcations, Springer (1969) New York.

69 P. Rabinowitz, Some global results for nonlinear eigenvalue problems. J. Func. Anal. 7 (1971) 487-513.

70 H. M. Rodrigues and A. L. Vanderbawhede, Symmetric perturbations of nonlinear equations: symmetry of solutions. JNA-TMA. To appear.

71 D. Sather, Branching of solutions of an equation in Hilbert space. Arch. Rat. Mech. Anal. 36 (1970) 47-64.

72 D. Sather, Branching of solutions of nonlinear equations, Rocky Mountain J. Math. 3 (1973) 203-250.

73 D. Sather, Branching and stability for nonlinear shells. Proc. IUTAM/IMU Symp. on Appl. and Methods of Funct. Anal. to Problems of Mech., Marseille, Sept 1-6, 1975.

74 D. H. Sattinger, Group representation theory and branch points of nonlinear functional equations. SIAM J. Math. Anal. (to appear).

75 D. Sattinger, Topics in Stability and Bifurcation Theory. Lecture Notes in Math. 309 (1973) Springer-Verlag.

76 M. Shearer, Small solutions of a nonlinear equation in Banach space for a degenerate case. Proc. Roy. Soc. Edinburgh, Ser. A. (to appear).

77 M. J. Sewell, Some mechanical examples of catastrophe theory. Bull. Inst. Math. and Appl. 12 (1976) 163-172.

78 I. Stakgold, Branching of solutions of nonlinear equations. SIAM Review 13 (1971) 289-332.

79 F. Takens, Unfoldings of certain singularities of vector fields: generalized Hopf bifurcations. J. Differential Eqns. 14 (1973) 476-493.

80 F. Takens, Singularities of functions and vector fields. Nieuw Arch. voor Wisk, 3, XX (1972) 107-130.

81 F. Takens, Singularities of vector fields. Pub. I.H.E.S., 43 (1974) 47-100.

82 F. Takens, Forced Oscillations and Bifurcations. In Applications of Global Analysis. Comm. 3 of the Math. Inst. Rikjsuniversitat Utrecht (1974).

83 R. Thom, Structural Stability and Morphogenesis. Benjamin (1975). Translated from French.

84 R. Thom, Topological methods in biology. Top. 8 (1968) 313-335.

85 R. Thom and E. C. Zeeman, Bibliography on catastrophe theory. Dynamical Systems - Warwick (1974) 390-401. Lecture Notes in Math. Vol. 468, Springer-Verlag.

86 J. M. T. Thompson and G. W. Hunt, A General Theory of elastic Stability. Wiley-Interscience (1973).

87 M. M. Vainberg and V. A. Trenogin, The method of Liapunov and Schmidt in the theory of nonlinear differential equations and their further development. Russ. Math. Survey 17 (1962) 1-60.

88 M. M. Vainberg and V. A. Trenogin, Theory of Branching of Solutions of Nonlinear Equations. Noordhoff (1974).

89 A. Vanderbauwhede, Generic and non-generic bifurcation for the von Kármán equations. J. Math. Anal. Appl. (to appear).

90 G. Wasserman, Stability of Unfoldings. Lecture Notes in Math. Vol. 393 (1974) Springer-Verlag.

91 D. Westreich, Banach space bifurcation theory. Trans. A.M.S. 171 (1973) 135-156.

Professor J. K. Hale, Lefschetz Center for Dynamical Systems,
Brown University, Providence, Rhode Island 02912, U.S.A.

J L ERICKSEN

On the formulation of St.-Venant's problem

St.-Venant is perhaps best known for his researches on the inverse and semi-inverse methods in linear elasticity theory, particularly as they apply to prisms loaded only at or near the ends. Realizing that, in practise, the distributions of tractions on the ends are not known in detail, and that some of the fine details seem not to matter much, he provided formats for calculating representative solutions. The rules for singling out one involve specifying some averages of the tractions. Thus, as we see it, St.-Venant's Problem is to reduce to mathematics the problem of singling out such representative solutions, to put it roughly. Experience indicates that this is a slippery problem for linear elasticity theory, still more so for the nonlinear theory. Here, our purpose is to sketch out some lines of thought which seem to be relevant and to shape up some mathematical questions deserving more study. Our treatment is more formalistic than rigorous; we only rough out the problems and concentrate on solutions associated with inverse and semi-inverse methods. It is a matter of experience, both empirical and theoretical, that within some rather poorly understood limits, these are rather representative. It seems not to have been noticed that they correspond to extremals of a certain functional which, evidently, must play an important role in the general problem. Thus, in part, our task is to make this clear. Sternberg and Knowles [1], dealing with linear theory, have shown that, with quite stringent side conditions, the St.-Venant solutions for tension, torsion and bending minimize the energy functional, a natural one to consider. Our side conditions are more natural, but our functional arises from reasoning more mathematical than physical.

The St.-Venant solutions for tension, torsion and bending are based on certain assumptions of symmetry concerning the material, shape of body and of its deformations. Our semi-inverse assumptions presume similar symmetry. It is fairly easy to relax traditional restrictions and to cover a wider range of physical problems, so we do this.

Of course, there is the point that, if one solution is representative of a set, so are the others. If we single out one, we do so because it is easier to analyze or, more importantly, is based on simple concepts. Said differently, aesthetics plays some role. Most of us might agree that, of St.-Venant's semi-inverse methods, that for flexure is least attractive, most difficult to extend to nonlinear elasticity theory. It is not terribly difficult to analyze, but it seems not to spring from simple concepts.

To avoid distraction from the main lines of thought, we exclude constrained materials. It is not too hard to adapt analyses to incompressible materials, to materials which are inextensible in certain directions, etc.

2 BASIC EQUATIONS

We presume that the reader has some familiarity with the elements of the non-linear theory of elastostatics. We take as basic Kirchhoff's form of the basic equations, with zero body force, viz.,

$$\left(\frac{\partial W}{\partial x^i{}_{,K}}\right)_{,K} = 0, \tag{2.1}$$

where, with x^i and X^K representing rectangular Cartesian spatial and material coordinates, respectively, the deformation is described by invertible mappings of the form

$$x^i = x^i\ (X^K). \tag{2.2}$$

Also, the strain energy function W is given by a constitutive equation of the form

$$W = W(x^i{}_{,K};\ X^L), \tag{2.3}$$

the energy E associated with a material region B being given by

$$E = \int_B W\ dX^1 dX^2 dX^3. \tag{2.4}$$

If a part of ∂B is given parametrically by $X^K = \dot{X}^K\ (u^1,\ u^2)$, the outward directed element of area is given by

$$dS_K = \pm\ \varepsilon_{KLM}\ \frac{\partial X^L}{\partial u^1}\ \frac{\partial X^M}{\partial u^2}\ du^1\ du^2, \tag{2.5}$$

the sign depending on the relative orientations of the material and surface coordinates. The force acting on such an element is given by

$$\frac{\partial W}{\partial x^{i}_{,k}} dS_K = f_i du^1 du^2, \tag{2.6}$$

the moment about $\underset{\sim}{x} = \underset{\sim}{x}_o$ by

$$\varepsilon_{ijk} (x^j - x_o^{\ j}) f_k \, du^1 \, du^2. \tag{2.7}$$

That the material coordinates are rectangular Cartesian is more in our minds than in the analysis. If we follow the above rules, we can equally well regard them as curvilinear, with commas denoting partial derivatives. The need for covariant derivatives or abstract equivalents encountered in common treatments comes about because of introduction of extraneous factors. If we insist that W always be interpretable as energy per unit Euclidean volume, we will be led to insert an additional factor in (2.4) and to modify (2.5). We evade these practices, sticking with the simpler rules. It is of more importance that the x^i be rectangular Cartesian. This stems from the common requirements of Galilean invariance,

$$W(\underset{\sim}{x}_{,K}; X^L) = W(\underset{\sim}{R} \, \underset{\sim}{x}_{,K}; \ X^L), \tag{2.8}$$

where $\underset{\sim}{R}$ is any rotation matrix,

$$\underset{\sim}{R}^{-1} = \underset{\sim}{R}^T , \quad \det \underset{\sim}{R} = 1. \tag{2.9}$$

One could, of course, describe the equivalent using say polar coordinates, but not in the same way nor as simply. Here, as in the following, we shift to more direct notation whenever this will not cause confusion, replacing $x^i_{\ ,K}$ by $\underset{\sim}{x}_{,K}$, etc.

Partly to elaborate these matters and partly to introduce formalism attending generalized coordinates to be used, we sketch out pertinent transformation theory. Additional discussion and coverage of more general transformations is given by Ericksen [2,3]. Consider invertible transformations of the form

$$\begin{cases} X^K = X^K (Y^\Gamma), \\ x^i = g^i (y^\alpha; Y^\Gamma), \end{cases} \tag{2.10}$$

where the different types of indices all take on the values 1,2,3. Then, by composition of mappings, (2.2) converts to relations of the form

$$y^\alpha = y^\alpha (Y^\Gamma) \tag{2.11}$$

and, by the chain rule

$$x^i{}_{,K} = \left(\frac{\partial g^i}{\partial y^\alpha} y^\alpha{}_{,\Gamma} + \frac{\partial g^i}{\partial Y^\Gamma} \right) Y^\Gamma{}_{,K}. \tag{2.12}$$

If we set

$$\begin{cases} J = |\det X^K{}_{,\Gamma}| \\ \bar{W} = WJ, \end{cases} \tag{2.13}$$

it is easy to see that $\bar{W}$ is expressible in the form

$$\bar{W} = \bar{W} (y^\alpha; y^\beta{}_{,\Gamma}; Y^\Delta). \tag{2.14}$$

It is a bit more laborious, but straightforward, to show that

$$\left(\frac{\partial \bar{W}}{\partial y^\alpha{}_{,\Gamma}} \right)_{,\Gamma} - \frac{\partial \bar{W}}{\partial y^\alpha} = J \frac{\partial g^i}{\partial y^\alpha} \left(\frac{\partial W}{\partial x^i{}_{,K}} \right)_{,K}. \tag{2.15}$$

Using (2.5) with X^K replaced by Y^Γ to define dS_Γ, we find that

$$\begin{cases} dS_\Gamma = J^{-1} X^K{}_{,\Gamma}\, dS_K, \\ \dfrac{\partial \bar{W}}{\partial y^\alpha{}_{,\Gamma}}\, dS_\Gamma = \dfrac{\partial g^i}{\partial y^\alpha} \dfrac{\partial W}{\partial x^i{}_{,K}}\, dS_K. \end{cases} \tag{2.16}$$

Thus, for example, (2.1) generates the equivalent Euler-Lagrange equations

$$\left(\frac{\partial \bar{W}}{\partial y^\alpha{}_{,\Gamma}}\right)_{,\Gamma} - \frac{\partial \bar{W}}{\partial y^\alpha} = 0, \tag{2.17}$$

while (2.4) becomes

$$\bar{E} = \int \bar{W}\, dY^1\, dY^2\, dY^3.$$

Perhaps this suffices to make clear the earlier remarks. With this explained, we henceforth assume that the X^K are either Cartesian or curvilinear, as convenient, and, in $(2.10)_1$, deal only with the special case where $\underset{\sim}{Y} = \underset{\sim}{X}$.

3 KINEMATICS

To begin with, let us consider a body whose natural state, or configuration of lowest energy, is of the traditional prismatic form. It can be thought of as generated by a cross-section, a plane region, referred to plane coordinates X^α, $X^\alpha \in D$, which could be Cartesian or curvilinear. The body is generated by translating this region along its normals. If X^3 denotes distance from the section, measured along the normals, and the length of the prism is L, then it occupies the region

$$\Delta = D \times [0,L]. \tag{3.1}$$

Geometrically, the various cross-sections X^3 = const. are, of course, congruent. If the material properties do not change as we move along the normals to the cross-sections, we should have

$$W = W(\underset{\sim}{x}_{,K};\, X^\alpha), \tag{3.2}$$

i.e., W should not depend explicitly on X^3. The body might be homogeneous, in which case W would not depend explicitly on X^α, if these are Cartesian coordinates. Such homogeneity, assumed by St.-Venant and many later workers, is not essential for our purposes. For a drawn wire, one might well assume that the properties vary with radius, for example. Assumptions of isotropy or transverse isotropy might simplify some analyses but, like St.-Venant, we allow anisotropic materials.

For combined tension and torsion, or bending, the traditional semi-inverse assumption is that the length may change and the cross-sections distort, but only in such a way that, both geometrically and mechanically, one cross-section remains indistinguishable from any other. One way of putting it is that, to within a rigid rotation, which can vary from one particle to another, the deformation gradient $\underset{\sim}{x}_{,K}$ is independent of X^3. As is discussed by Ericksen [2], it is feasible to characterize such deformations. Let α and γ be constants, $\underset{\sim}{k}$ be the coordinates of a fixed point. Let $\underset{\sim}{e}$ be a constant unit vector,

$$\underset{\sim}{e} \cdot \underset{\sim}{e} = 1, \tag{3.3}$$

and let $\underset{\sim}{R} = \underset{\sim}{R}(X^3)$ be the rotation matrix with $\underset{\sim}{e}$ as axis, γX^3 as angle of rotation. One representation of it obtains as follows: let $\underset{\sim}{\Omega}$ be the skew matrix equivalent to $\underset{\sim}{e}$,

$$\begin{cases} \underset{\sim}{\Omega}^T = - \underset{\sim}{\Omega} \\ \underset{\sim}{\Omega}\, \underset{\sim}{v} = \underset{\sim}{e} \wedge \underset{\sim}{v} \quad \forall\, \underset{\sim}{v}. \end{cases} \tag{3.4}$$

We then have

$$\begin{cases} \underset{\sim}{R} = \exp(\gamma X^3 \underset{\sim}{\Omega}) = \underset{\sim}{1} + \gamma X^3 \underset{\sim}{\Omega} + \cdots \\ \underset{\sim}{R}(0) = \underset{\sim}{1}, \\ \underset{\sim}{R}\, \underset{\sim}{e} = \underset{\sim}{e} \\ \underset{\sim}{R}' = \gamma \underset{\sim}{R}\, \underset{\sim}{\Omega} = \gamma \underset{\sim}{\Omega}\, \underset{\sim}{R}. \end{cases} \tag{3.5}$$

The cross-section $X^3 = 0$ will become some surface. With X^α serving as surface coordinates, it will be described by

$$\underset{\sim}{x} = \underset{\sim}{y}(X^\alpha) \text{ when } X^3 = 0,\ X^\alpha \in D. \tag{3.6}$$

The mappings of interest are then given by

$$\begin{cases} \underset{\sim}{x} = \underset{\sim}{k} + \underset{\sim}{R}\,(\underset{\sim}{y} - \underset{\sim}{k} + \alpha\, X^3\, \underset{\sim}{e}) \\ \quad = \underset{\sim}{k} + \underset{\sim}{R}\,(\underset{\sim}{y} - \underset{\sim}{k}) + \alpha\, X^3\, \underset{\sim}{e}. \end{cases} \tag{3.7}$$

In these terms, the natural state is given by

$$\begin{cases} \alpha = 1,\ \gamma = 0,\ \underset{\sim}{e} = (0,0,1) \\ \underset{\sim}{y}\,(X^\alpha) = (X^1,\ X^2,\ 0), \end{cases} \tag{3.8}$$

when the X^α are taken as rectangular Cartesian coordinates. When $\alpha\,\gamma \neq 0$, the initially straight material lines X^α = const. become circular helices, all centered on the line through $\underset{\sim}{x} = \underset{\sim}{k}$, parallel to $\underset{\sim}{e}$. We call this line the <u>twist axis</u>. By tradition, torsion refers to the case where it is the line of centroids of the cross-sections, thus lying within the outer confines of the body. If the twist axis lies well outside the body, the deformed body has a shape more like that of a helical spring. If one then goes to the limiting case where $\alpha = 0$, the helices degenerate to circles and one has typical bending deformations. If, on the other hand, $\gamma = 0$, the twisting disappears, the helices degenerating to straight lines. One then has deformations more like those encountered in studies of simple tension, plain strain, etc. Formally, some such deformations will have different parts of the body occupying the same parts of space, so there are non-trivial topological problems associated with global invertibility. We ignore these. We note that, from (3.5), (3.6) and (3.7),

$$\begin{cases} \underset{\sim}{x}_{,\alpha} = \underset{\sim}{R}\, \underset{\sim}{y}_{,\alpha}, \\ \underset{\sim}{x}_{,3} = \underset{\sim}{R}\, \underset{\sim}{z}, \\ \underset{\sim}{z} \ \ = \gamma\, \underset{\sim}{e} \wedge (\underset{\sim}{y} - \underset{\sim}{k}) + \alpha\, \underset{\sim}{e}, \end{cases} \tag{3.9}$$

so the Jacobian test for local invertibility reads, say,

$$\underset{\sim}{x}_{,1} \wedge \underset{\sim}{x}_{,2} \cdot \underset{\sim}{x}_{,3} = \underset{\sim}{y}_{,1} \wedge \underset{\sim}{y}_{,2} \cdot \underset{\sim}{z} > 0. \tag{3.10}$$

Thus, for example, we cannot have $\alpha = \gamma = 0$.

The classical linear theory involves, among other things, an approximation such as

$$\underset{\sim}{R} \cong \underset{\sim}{1} + \gamma X^3 \underset{\sim}{\Omega}, \tag{3.11}$$

a slight variant, which makes sense only if the total twist angle is small,

$$|\gamma| L \ll 1, \tag{3.12}$$

a rather restrictive limitation. For the semi-inverse solutions, it is quite easy to do a similar linearization which lets $\underset{\sim}{R}$ remain a finite rotation. If one looks at Love's [4] critique of the Kirchoff theory of thin rods, one will see one of the older attempts to cope with such corrections. Of course, the aim there is to compare thin rod theory with three-dimensional theory, for cases where $|\gamma|d$ and d/L are small, d being the diameter of the cross-section. In this, Love makes rather heavy use of the St.-Venant solutions. Thoughtful reconsideration and analyses of such thin body approximations should provide useful guidelines for those interested in modern theories of thin elastic rods and shells. We have sought a format which, hopefully, will facilitate such analyses.

We began by assuming a body with natural state of prismatic form, interpreting X^3 as a Cartesian coordinate. Some of the ensuing analyses go through equally well for more general bodies. Each of the deformations (3.7) suggests a type of body, together with a coordinatization, which could serve to describe a natural state, such that W does not depend explicitly on X^3. Of course, (3.8) must then be modified in an appropriate way. Thus, the mathematical theory of semi-inverse methods for spiral columns or helical springs is quite like that of prisms.

4 EQUILIBRIUM EQUATIONS

Since, for the configurations considered, all cross-sections are alike, it is intuitively evident that all should be in equilibrium if one is. Of course, such configurations can be expected to obtain only for rather special types of loadings. Following St.-Venant and using (3.1), we require that the lateral surface be free,

$$\frac{\partial W}{\partial \underset{\sim}{x}_{,K}} \, dS_K = \underset{\sim}{0} \quad \text{on } \partial D \times [0,L]. \tag{4.1}$$

It is, of course, feasible to admit non-zero tractions of special types, generalizing problems associated with the names of Almansi and Hamel in linear elasticity theory. Such problems are briefly discussed by Ericksen [2], but we here stick to the simpler case.

On any cross-section X^3 = const., we can use X^α as surface coordinates. Then, with dS_K pointing in the direction of increasing X^3, (2.5) gives

$$dS_1 = dS_2 = 0 \ , \ dS_3 = dS, \tag{4.2}$$

where

$$dS = dX^1 dX^2, \tag{4.3}$$

may or may not represent the element of Euclidean area, depending upon the interpretation of coordinates. Using (2.6) and (2.7), the resultant force acting on X^3 = const. is

$$\underset{\sim}{F}\,(X^3) = \int_D \frac{\partial W}{\partial \underset{\sim}{x}_{,3}} \, dS, \tag{4.4}$$

while

$$\underset{\sim}{M}(X^3) = \int_D (\underset{\sim}{x} - \underset{\sim}{k}) \wedge \frac{\partial W}{\partial \underset{\sim}{x}_{,3}} \, dS \tag{4.5}$$

gives the resultant moment about $\underset{\sim}{x} = \underset{\sim}{k}$. According to St.-Venant, it is these averages, rather than the fine details of loading, which are of primary interest. It is not our purpose to explore the validity of using the semi-inverse solutions when the actual loadings are somewhat different, but to concentrate on the special solutions themselves. However, §6 covers results having some bearing on this issue.

We now turn to the question of reducing the analysis of a problem to one for a cross-section, say $X^3 = 0$. First, ignore (3.6) and view (3.7) as a generalized coordinate transformation of the type (2.10). In the new coordinates, a general deformation would be described by relations of the form (2.11), $\underset{\sim}{y}$ depending on X^3 as well as X^α. Then (3.9) would be replaced by relations only slightly different, viz.,

$$\begin{cases} \underset{\sim}{x}_{,\alpha} = \underset{\sim}{R}\, \underset{\sim}{y}_{,\alpha} , \\ \underset{\sim}{x}_{,3} = \underset{\sim}{R}\, \underset{\sim}{z} , \\ \underset{\sim}{z} = \gamma\, \underset{\sim}{e} \wedge (\underset{\sim}{y} - \underset{\sim}{k}) + \alpha\, \underset{\sim}{e} + \underset{\sim}{y}_{,3} . \end{cases} \tag{4.6}$$

Using (2.13) and (2.8), we have

$$\begin{aligned} \bar{W}(\underset{\sim}{y}\,;\, \underset{\sim}{y}_{,\alpha}\,;\, \underset{\sim}{y}_{,3},\, X^\beta) &= W(\underset{\sim}{x}_{,\alpha}\,;\, \underset{\sim}{x}_{,3}\,;\, X^\beta) \\ &= W(\underset{\sim}{y}_{,\alpha}\,;\, \underset{\sim}{z}\,;\, X^\beta), \end{aligned} \tag{4.7}$$

so, like W, $\bar{W}$ will not depend explicitly on X^3. In fact, since $\underset{\sim}{z}$ does not depend explicitly on any material coordinate, $\bar{W}$ will be independent of X^2 if W is, etc. From (2.17), the new equilibrium equations are

$$\left(\frac{\partial \bar{W}}{\partial \underset{\sim}{y}_{,K}}\right)_{,K} - \frac{\partial \bar{W}}{\partial \underset{\sim}{y}} = \underset{\sim}{0}. \tag{4.8}$$

Thus, if we now make the semi-inverse assumption that $\underset{\sim}{y}_{,3} = \underset{\sim}{0}$, we get equations independent of X^3, viz.,

$$\left(\frac{\partial \bar{W}}{\partial \underset{\sim}{y}_{,\alpha}}\right)_{,\alpha} - \frac{\partial \bar{W}}{\partial \underset{\sim}{y}} = \underset{\sim}{0}, \tag{4.9}$$

interpretable as equations for the cross-section $X^3 = 0$.

The lateral surface can be described by giving ∂D a parametric description of the form $X^\alpha = X^\alpha(u)$, then using u and X^3 as surface coordinates on $\partial D \times [0,L]$. Then, using (2.5) and (4.1), we obtain the reduced form

$$\frac{\partial \bar{W}}{\partial \underset{\sim}{y}_{,\alpha}} ds_\alpha = \underset{\sim}{0}, \tag{4.10}$$

where

$$ds_\alpha = \pm \, \varepsilon_{\alpha\beta} \frac{\partial X^\beta}{du} du, \tag{4.11}$$

describes the integration element on ∂D. Formally, if we vary the energy $\hat{E}$ of the cross-section,

$$\hat{E} = \int_D \bar{W} \, dS, \tag{4.12}$$

without restricting $\delta \underset{\sim}{y}$ on ∂D, we obtain, as conditions for its extremals, the Euler-Lagrange equations (4.9) and natural boundary conditions (4.10). Thus, for the cross-section itself, one can use ideas of minimum energy to rank solutions when there is multiplicity, or use variational methods to study existence of solutions, etc.

From (3.5), (4.4), (4.6) and (4.7), we find that

$$\underset{\sim}{F}(X^3) = \underset{\sim}{R}\int_D \frac{\partial W}{\partial \underset{\sim}{z}}\, dS = \underset{\sim}{R}\,\underset{\sim}{F}(0). \tag{4.13}$$

On the other hand, with the equivalent of the free boundary condition (4.1) and equilibrium equations (2.1) satisfied, we must have balance of forces, implying that $\underset{\sim}{F}(X^3) = \underset{\sim}{F}(0)$. If $\gamma \neq 0$, so that $\underset{\sim}{R} \neq \underset{\sim}{1}$, $\underset{\sim}{F}$ must have the direction of the axis of rotation,

$$\underset{\sim}{F} = f\,\underset{\sim}{e}, \tag{4.14}$$

$$f = \underset{\sim}{e} \cdot \int_D \frac{\partial W}{\partial \underset{\sim}{x}_{,3}}\, dS = \underset{\sim}{e} \cdot \int_D \frac{\partial W}{\partial \underset{\sim}{z}}\, dS = \frac{\partial \hat{E}}{\partial \alpha} = \text{const.} \tag{4.15}$$

Similar consideration of (4.5), again assuming $\gamma \neq 0$, yields

$$\underset{\sim}{M}(X^3) = \underset{\sim}{M}(0) = m\,\underset{\sim}{e}, \tag{4.16}$$

$$m = \underset{\sim}{e} \cdot \int_D (\underset{\sim}{y} - \underset{\sim}{k}) \wedge \frac{\partial W}{\partial \underset{\sim}{z}}\, dS = \frac{\partial \hat{E}}{\partial \gamma} = \text{const.} \tag{4.17}$$

Actually, provided that $\underset{\sim}{F} \neq \underset{\sim}{0}$, (4.14) - (4.17) still apply when $\gamma = 0$. The condition that the Cauchy stress tensor be symmetric, which is implied by (2.8), reads

$$\underset{\sim}{x}_{,K} \wedge \frac{\partial W}{\partial \underset{\sim}{x}_{,K}} = \underset{\sim}{0}, \tag{4.18}$$

so, with (3.9) and $\gamma = 0$, we have

$$\alpha\,\underset{\sim}{e} \wedge \int_D \frac{\partial W}{\partial \underset{\sim}{x}_{,3}}\, dS = -\int_D \underset{\sim}{x}_{,\alpha} \wedge \frac{\partial W}{\partial \underset{\sim}{x}_{,\alpha}}\, dS.$$

With the null boundary conditions and equilibrium equations satisfied, it is easily shown that the right member vanishes, which implies (4.14). Now $\underset{\sim}{k}$ cancels out of (3.7). However, it is a small exercise to show that we can choose it so (4.16) holds. If $\gamma = \underset{\sim}{F} = 0$, exceptions to (4.16) can occur. For present purposes, such cases are not of major interest, so we exclude them.

For a given body, the naive expectation is that solving the cross-section problem will produce solutions depending rather smoothly on the parameters α and γ, with

$$f = f(\alpha,\gamma) \ , \ m = m(\alpha,\gamma), \tag{4.19}$$

being invertible functions of the parameters. With (4.15) and (4.17) in mind, one might equally well estimate common prejudice as favouring the notion that $\hat{E}$ is a convex function of α and γ. In our view, contrary cases are not devoid of physical interest. Nevertheless, it would be of interest to know what must be presumed of W, etc., to rigorously establish such smoothness and convexity conditions. It is perhaps worth remarking that some breakdown of existence of solutions, etc., for the cross-section problem might only mean that the body prefers to deform in a manner incompatible with the semi-inverse assumptions.

It is easily shown that certain transformations convert solutions of (4.9) and (4.10) into solutions. Bearing in mind (2.8), (4.6) and (4.7), we can add any constant vector to $\underset{\sim}{y}$ and $\underset{\sim}{k}$ or, if it is parallel to $\underset{\sim}{e}$, to either one by itself. We can also apply to $\underset{\sim}{y} - \underset{\sim}{k}$ any constant rotation with $\underset{\sim}{e}$ as axis. Thus one can introduce any convenient normalizations to eliminate the implied ambiguities, or regard as equivalent solutions so related. From the old theorems of Emmy Noether [5], we expect this invariance under continuous groups to yield two conservation laws. They are

$$\underset{\sim}{e} \cdot \left(\frac{\partial \bar{W}}{\partial \underset{\sim}{y}_{,\alpha}} \right)_{,\alpha} = 0, \tag{4.20}$$

and

$$\underset{\sim}{e} \cdot \left(\underset{\sim}{y} \wedge \frac{\partial \bar{W}}{\partial \underset{\sim}{y}_{,\alpha}} \right)_{,\alpha} = 0. \tag{4.21}$$

We note that such translations and rotations do not affect the values of f and m.

If one wishes to replace the neo-Cartesian coordinates $\underset{\sim}{y}$ by some which are curvilinear, one can use the general transformation scheme to obtain the new Euler-Lagrange equations, etc. We have no need for this.

5 TORSION

Perhaps because of a perverted sense of humor, we more enjoy an old problem if we can give it a new twist, and where better to do this than in the theory of torsion of prisms. Introduction of a measure of area or mass will enable us to define a center $\underset{\sim}{c}\,(X^3)$ of any cross-section X^3 = const., as

$$\underset{\sim}{c}\,(X^3) = \int_D \underset{\sim}{x}\,\mu\,(X^\alpha)\,dS \Big/ \int_D \mu\,dS, \tag{5.1}$$

where $\mu > 0$ generates the measure selected. For combined tension and torsion, $\underset{\sim}{F}$ and $\underset{\sim}{M}$ will be in the direction $\underset{\sim}{e}$, which we regard as specified. For such problems, a likely side condition is

$$\underset{\sim}{c}\,(L) - \underset{\sim}{c}\,(0) = \beta\,\underset{\sim}{e}. \tag{5.2}$$

We think of the body loaded by a rather hard device, able to hold the mean positions of the end where we want them, and to keep the overall twist angle where we want it. Said differently, β and γ are control parameters and there is no real loss of generality in fixing conditions at an end so that

$$\underset{\sim}{c}\,(0) = \underset{\sim}{0}. \tag{5.3}$$

Using (3.7), (5.1), (5.2) and (5.3), we then calculate that

$$\underset{\sim}{c}\,(L) = \beta\,\underset{\sim}{e} = \underset{\sim}{k} - \underset{\sim}{R}\,(L)\,\underset{\sim}{k} + \alpha\,L\,\underset{\sim}{e}\,. \tag{5.4}$$

Dotting this with $\underset{\sim}{e}$ gives

$$\alpha = \beta/L, \tag{5.5}$$

so α can replace β as a control parameter, L being thought of as fixed. Obviously, for torsion, we want there to be some rotation so, from (5.4) and (5.5), we should have $\underset{\sim}{k}$ parallel to $\underset{\sim}{e}$, the axis of rotation which means that there is no loss in generality in equating it to zero. The twist axis then coincides with the line of centers. On second thought, we can have $\underset{\sim}{R}(L) = \underset{\sim}{1}$, with $\underset{\sim}{R} \not\equiv \underset{\sim}{1}$ if the total twist angle γL is a multiple of 2π, i.e.,

$$\gamma = 2n\,\pi/L \quad , \quad n = \pm 1, \pm 2, \cdots. \tag{5.6}$$

When these critical conditions obtain, (5.4) imposes no conditions on $\underset{\sim}{k}$, so the twist axis could, say, jump outside the body, the line of centers becoming a circular helix. In severely twisted rubber bands or ribbons, we have all seen changes in the mode of deformation somewhat like this. This would suggest that, for n sufficiently large, there might well be some such shift in the twist axis, tending to give a lower value of the cross-section energy $\hat{E}$. If so, this is perhaps best regarded as an indication that we have gone past the limits of applicability of the semi-inverse solutions. Of course, we glimpse only part of the picture by studying these special solutions alone. On the other hand, it is not entirely trivial to mathematize the basic physical problem in a way which allows for more general solutions.

In the theory of the elastic rods, it is customary to think of the cross-section as simply connected and of some rather simple shape. For simplicity, think of the prism as homogeneous, take X^α as Cartesian coordinates with origin at the center. With $\mu = 1$, this will be the traditional center of mass.

We attempt to formalize some of the traditional prejudices, as they apply to the torsion problems. From (4.6), (4.7) and (4.12), we have

$$\hat{E} = \int_D W[\underset{\sim}{y}_{,\alpha} \; ; \; \gamma \underset{\sim}{e} \wedge (\underset{\sim}{y} - \underset{\sim}{k}) + \alpha \underset{\sim}{e}] \, dS. \tag{5.7}$$

In the thin body approximation, we commonly intend that the diameter of D is small compared to L, so we fix L and consider shrinking D. Ignoring the previously discussed subtlety, and bearing in mind the translational invariance of W, we set

$$\underset{\sim}{k} = \underset{\sim}{c}(0) = \underset{\sim}{0}, \tag{5.8}$$

and consider the material and parameters α, γ and $\underset{\sim}{e}$ as fixed.
Now put

$$\begin{cases} X^\alpha = \varepsilon \, Z^\alpha, \\ \underset{\sim}{y} = \varepsilon \, \underset{\sim}{p} \, , \\ \bar{E} = \hat{E}/\varepsilon^2, \end{cases} \tag{5.9}$$

letting $\bar{D}$ be the image of D under this map. Then

$$\bar{E} = \int_{\bar{D}} W \left(\frac{\partial \underset{\sim}{p}}{\partial Z^\alpha} \; ; \gamma \, \varepsilon \, \underset{\sim}{e} \wedge \underset{\sim}{p} + \alpha \, \underset{\sim}{e} \right) dZ^1 \, dZ^2. \tag{5.10}$$

Then, if we fix $\bar{D}$ and let ε get small, we will be shrinking D by a similarity transformation, approaching the centroid as $\varepsilon \longrightarrow 0$. Clearly, this is energetically equivalent to fixing D and replacing γ by $\gamma \, \varepsilon$ in (5.7), i.e., to fixing D and letting the twist per unit length become small, a type of perturbation which is familiar in the theory of small deformations superposed on large. There is, however, a rather subtle difference, for we have factored out $\underset{\sim}{R}$. For the former version, γ and γL are fixed, and γL can be large, so the rotation $\underset{\sim}{R}$ remains finite as $\varepsilon \longrightarrow 0$. As the latter version is encountered in practice, $\underset{\sim}{R}$ would become infinitesimal as the twist per unit length becomes small, $\underset{\sim}{R} \longrightarrow \underset{\sim}{1}$ in the limit. Of course, one could let L grow as ε^{-1} and force equivalence of the two perturbations, but this has not been the practice. Taking the former version, in the limit $\varepsilon = 0$, a likely solution has $\partial p/\partial Z^{\alpha}$ = const., satisfying

$$\frac{\partial W}{\partial(\partial \underset{\sim}{p}/\partial Z^{\alpha})} = \underset{\sim}{0}. \tag{5.11}$$

At $X^3 = 0$, $\underset{\sim}{R} = \underset{\sim}{1}$ and, since now $\underset{\sim}{x}_{,3} = \underset{\sim}{R}\, \alpha\, \underset{\sim}{e} = \alpha\, \underset{\sim}{e}$, this is everywhere constant. The solution is much like a simple tension solution, except that the rotation influences $\underset{\sim}{x}_{,\alpha} = \underset{\sim}{R}\, \partial \underset{\sim}{p}/\partial Z^{\alpha}$. Perhaps these naive and cursory considerations suffice to make clear that this is not the usual perturbation theory for small twist superposed on finite tension. However, it seems likely that one could evolve a straightforward formal asymptotic theory and, perhaps, give it a rigorous treatment. With slight modification, such analyses could cover the other semi-inverse solutions, and it would seem feasible to similarly treat cylindrical shells, by allowing doubly connected cross-sections. In our view, this is the most promising place to begin to upgrade the theories of slender bodies.

For bodies of the general type here considered, loaded only near the ends, it is a matter of experience that solutions of the type previously discussed are sometimes, but not always representative. There are some points of similarity between these and other possible solutions. Also, some factors which might lead to a better understanding of the situation seem not to be appreciated.

In concept, reference configurations of the type here considered, together with their coordinatizations, equip a body with identifiable material cross-sections X^3 = const. It is only that they need not all deform in like manner. If the body is loaded near the ends, we can fix a subsection $0 \leqslant X^3 \leqslant L$ with free lateral boundary which, for theoretical purposes, can here serve as the body. The resultant force on a cross-section is still given by

$$\underset{\sim}{F}(X^3) = \int_D \frac{\partial W}{\partial \underset{\sim}{x}_{,3}} \, dS,$$

and, in equilibrium, we must have

$$\underset{\sim}{F}(X^3) = \underset{\sim}{F}(0) \,, \quad 0 \leqslant X^3 \leqslant L.$$

Similarly, the resultant moment about $\underset{\sim}{x} = \underset{\sim}{x}_o$ is given by

$$\underset{\sim}{M}(X^3) = \int_D (\underset{\sim}{x} - \underset{\sim}{x}_o) \wedge \frac{\partial W}{\partial \underset{\sim}{x}_{,3}} \, dS$$

and, in equilibrium, satisfies

$$\underset{\sim}{M}(X^3) = \underset{\sim}{M}(0).$$

If $\underset{\sim}{F} = \underset{\sim}{0}$, $\underset{\sim}{M}$ does not depend on the choice of $\underset{\sim}{x}_o$. If $\underset{\sim}{F} \neq \underset{\sim}{0}$, it is easy to show that there are some choices of $\underset{\sim}{x}_o$ for which $\underset{\sim}{M}$ is parallel to $\underset{\sim}{F}$. Choosing notation to emphasize similarities, we let $\underset{\sim}{x}_o = \underset{\sim}{k}$ denote such a choice. We then have

$$\begin{cases} \underset{\sim}{F} = f\,\underset{\sim}{e}, \\ f = \underset{\sim}{e} \cdot \displaystyle\int_D \frac{\partial W}{\partial \underset{\sim}{x}_{,3}}\, dS = \text{const.}, \end{cases} \tag{6.1}$$

$$\begin{cases} \underset{\sim}{M} = m\,\underset{\sim}{e}, \\ m = \underset{\sim}{e} \cdot \displaystyle\int_D (\underset{\sim}{x} - \underset{\sim}{k}) \wedge \frac{\partial W}{\partial \underset{\sim}{x}_{,3}}\, dS = \text{const.}, \end{cases} \tag{6.2}$$

where $\underset{\sim}{e}$ is now picked out as the common direction of $\underset{\sim}{F}$ and $\underset{\sim}{M}$, unless both vanish, in which case it is arbitrary. Thus we have analogues of (4.14)-(4.17). Clearly, we can replace $\underset{\sim}{k}$ by any point on the line through it, parallel to $\underset{\sim}{e}$, giving us an analogue of the twist axis. Here, the line becomes indeterminate if $\underset{\sim}{F} = \underset{\sim}{0}$, which could occur in torsion or bending. Of course, one might devise kinematic conditions to pin it down, and we did lean in this direction in the discussion of torsion. Since the values of f and m are the same for all cross-sections, we cannot expect two solutions to approach each other as we move away from the ends unless these values match or are at least very close. Of course, this point was recognized by St.-Venant, who suggested that we make the match precise. Gradually, workers became aware that something else must be said. For infinitesimal deformations, theorems like those established by Toupin [6] have provided some illumination. It does seem evident that any other averages over cross-sections, which are the same for all cross-sections, must similarly match, and at least one seems to have been overlooked. With W not depending explicitly on X^3, it is easily verified that, when (2.1) is satisfied,

$$W_{,3} - \left(\frac{\partial W}{\partial \underset{\sim}{x}_{,K}} \cdot \underset{\sim}{x}_{,3} \right)_{,K} = 0. \tag{6.3}$$

This is the conservation law which, Emmy Noether taught us, should follow from having W invariant under the one-parameter group $X^3 \longrightarrow X^3 + \text{const.}$ If we integrate this over a cross-section and use the condition that the lateral boundary is free, we find that

$$g'\ (X^3) = \int_D \left(\frac{\partial W}{\partial \underset{\sim}{x}_{,\alpha}} \cdot \underset{\sim}{x}_{,3} \right)_{,\alpha} dS = \int_{\partial D} \frac{\partial W}{\partial \underset{\sim}{x}_{,\alpha}} \cdot \underset{\sim}{x}_{,3}\ dS_\alpha = 0,$$

where

$$\begin{aligned} g\ (X^3) &= \int_D \left(W - \frac{\partial W}{\partial \underset{\sim}{x}_{,3}} \cdot \underset{\sim}{x}_{,3} \right) dS \\ &= g\ (0) \quad , \quad 0 \leqslant X^3 \leqslant L. \end{aligned} \tag{6.4}$$

Thus there are at least the three integrals to match and it does not appear that g is determined by f and m, in general. It seems difficult to exclude the possibility that still more such integrals exist but, for purposes of discussion, we ignore this. It would seem that the semi-inverse solutions involve too few parameters to enable matching of f, m and g. From this viewpoint, it thus seems odd that they seem representative as often as they do. Toupin's [6] analysis does suggest the relevance of some energy-like measure of the difference between solutions, but not this one. His decays with distance from the end, in exponential fashion. There is then some suggestion that, if we match forces and moments, two solutions will become very close far from the ends, provided they are never too far apart. This seems to have relevance only to bodies which are moderately slender, for whatever that is worth. In pondering these matters, we hit upon one analysis which seems to shed a bit of light, so we turn to it.

The special solutions discussed earlier do have a special status, as extremals of the functional g. Suppose that we fix $\underset{\sim}{k}$ and the values of $\underset{\sim}{F}$ and $\underset{\sim}{M}$ at $\bar{\underset{\sim}{F}}$ and $\bar{\underset{\sim}{M}}$, respectively, in a manner consistent with (6.1) and (6.2), for some choice of the direction $\underset{\sim}{e}$. Fix any cross-section X^3 = const. On it, consider the restriction of deformations and deformation gradients which are consistent with the free boundary condition

$$\frac{\partial W}{\partial \underset{\sim}{x}_{,\alpha}} ds_{\alpha} = 0 \quad \text{on } \partial D. \tag{6.5}$$

Actually, one can omit this restriction and deduce (6.5) as a natural boundary condition. Formally, we can eliminate the constraints on $\underset{\sim}{F}$ and $\underset{\sim}{M}$ and introduce Lagrange multipliers $\underset{\sim}{a}$ and $\underset{\sim}{b}$, replacing g by

$$h = g + \underset{\sim}{a} \cdot (\underset{\sim}{F} - \bar{\underset{\sim}{F}}) + \underset{\sim}{b} \cdot (\underset{\sim}{M} - \bar{\underset{\sim}{M}}). \tag{6.6}$$

Using (6.1), (6.2), (6.3) and (6.5), we find, by a routine calculation, that the Gateaux differential of h is given by

$$\begin{aligned} \delta h = \int_D \Bigg\{ &- \left[\left(\frac{\partial W}{\partial \underset{\sim}{x}_{,\alpha}} \right)_{,\alpha} + \underset{\sim}{b} \wedge \frac{\partial W}{\partial \underset{\sim}{x}_{,3}} \right] \cdot \delta \underset{\sim}{x} \\ &+ \left[\underset{\sim}{a} + \underset{\sim}{b} \wedge (\underset{\sim}{x} - \underset{\sim}{k}) - \underset{\sim}{x}_{,3} \right] \cdot \delta \left(\frac{\partial W}{\partial \underset{\sim}{x}_{,3}} \right) \Bigg\} dS \\ &+ (\underset{\sim}{F} - \bar{\underset{\sim}{F}}) \cdot \delta \underset{\sim}{a} + (\underset{\sim}{M} - \bar{\underset{\sim}{M}}) \cdot \delta \underset{\sim}{b}. \end{aligned} \tag{6.7}$$

Now, in the interior of D, $\underset{\sim}{x}$ and $\underset{\sim}{x}_{,3}$ can be varied independently, X^3 being fixed. Further,

$$\delta\left(\frac{\partial W}{\partial \underset{\sim}{x}_{,3}}\right) = \frac{\partial^2 W}{\partial \underset{\sim}{x}_{,3}\,\partial \underset{\sim}{x}_{,\alpha}}\,\delta \underset{\sim}{x}_{,\alpha} + \frac{\partial^2 W}{\partial \underset{\sim}{x}_{,3}\,\partial \underset{\sim}{x}_{,3}}\,\delta \underset{\sim}{x}_{,3},$$

so, provided that

$$\det\left(\frac{\partial^2 W}{\partial \underset{\sim}{x}_{,3}\,\partial \underset{\sim}{x}_{,3}}\right) \neq 0, \tag{6.8}$$

we can assign the left member and $\delta\underset{\sim}{x}$ on D and calculate $\delta\underset{\sim}{x}_{,3}$ to keep the equation balanced. Here, (6.8) is the condition that the cross-section not be a characteristic surface, a rather mild limitation for elastostatics. Accepting it, we can regard $\delta\underset{\sim}{x}$ and $\delta(\partial W/\partial\underset{\sim}{x}_{,3})$ as arbitrary. Then, by equating to zero δh, we get the condition that the force and moment constraints be satisfied and

$$\left(\frac{\partial W}{\partial \underset{\sim}{x}_{,\alpha}}\right)_{,\alpha} + \underset{\sim}{b} \wedge \frac{\partial W}{\partial \underset{\sim}{x}_{,3}} = \underset{\sim}{0} \quad \text{in D}, \tag{6.9}$$

$$\underset{\sim}{x}_{,3} = \underset{\sim}{b} \wedge (\underset{\sim}{x} - \underset{\sim}{k}) + \underset{\sim}{a} \quad \text{in D}. \tag{6.10}$$

Integration of (6.9) over D, together with (6.5), yields

$$\underset{\sim}{b} \wedge \underset{\sim}{\bar{F}} = \underset{\sim}{0}. \tag{6.11}$$

If we cross (6.9) with $\underset{\sim}{x} - \underset{\sim}{k}$, integrate over D, then use (4.18) and (6.5), we obtain

$$\int_D \left\{\underset{\sim}{x}_{,3} \wedge \frac{\partial W}{\partial \underset{\sim}{x}_{,3}} + (\underset{\sim}{x} - \underset{\sim}{k}) \wedge \left[\underset{\sim}{b} \wedge \frac{\partial W}{\partial \underset{\sim}{x}_{,3}}\right]\right\} dS = \underset{\sim}{0}.$$

If we now employ (6.10) and standard vector identities, we obtain

$$\underset{\sim}{b} \wedge \underset{\sim}{\bar{M}} + \underset{\sim}{a} \wedge \underset{\sim}{\bar{F}} = \underset{\sim}{0}. \tag{6.12}$$

If neither $\bar{\underset{\sim}{M}}$ nor $\bar{\underset{\sim}{F}}$ vanish, both being parallel to $\underset{\sim}{e}$, (6.11) and(6.12) require that $\underset{\sim}{a}$ and $\underset{\sim}{b}$ both be parallel to $\underset{\sim}{e}$, so we can write

$$\underset{\sim}{a} = \alpha \underset{\sim}{e}, \quad \underset{\sim}{b} = \gamma \underset{\sim}{e}. \tag{6.13}$$

Similarly, (6.12) obtains if $\bar{\underset{\sim}{M}} = \underset{\sim}{0} \neq \bar{\underset{\sim}{F}}$. If $\bar{\underset{\sim}{F}} = \underset{\sim}{0} \neq \bar{\underset{\sim}{M}}$, (6.12) requires that $\underset{\sim}{b}$ be parallel to $\underset{\sim}{e}$, so $(6.13)_2$ applies. Also, changing $\underset{\sim}{k}$ then does not change the value of $\bar{\underset{\sim}{M}}$. If $\gamma \neq 0$, we can write

$$\underset{\sim}{a} = \alpha \underset{\sim}{e} + \gamma \underset{\sim}{e} \wedge \underset{\sim}{\ell},$$

resolving $\underset{\sim}{a}$ into parts parallel and perpendicular to $\underset{\sim}{e}$. We can then replace $\underset{\sim}{k}$ by $\underset{\sim}{k} - \underset{\sim}{\ell}$. Bearing in mind (6.10), this effectively gives (6.13). If $\bar{\underset{\sim}{F}} = \bar{\underset{\sim}{M}} = \underset{\sim}{0}$, $\underset{\sim}{e}$ is indeterminate, and our conditions do not force $\underset{\sim}{a}$ to be parallel to $\underset{\sim}{b}$. If not, we can take $\underset{\sim}{e}$ as the direction of $\underset{\sim}{b}$ and adjust $\underset{\sim}{k}$ as above to get (6.13). This leaves a possible exception,

$$\bar{\underset{\sim}{F}} = \underset{\sim}{0} \neq \bar{\underset{\sim}{M}}, \ \underset{\sim}{b} = \underset{\sim}{0}, \tag{6.14}$$

when $\underset{\sim}{a}$ might not be parallel to $\underset{\sim}{e}$. We restrict our attention to the typical case (6.13), leaving to the reader the perusal of (6.14).

If (6.9) and (6.10) obtain on every cross-section then, nominally, α and γ might depend on X^3. Introduce functions $\sigma(X^3)$ and $\tau(X^3)$ such that

$$\alpha = \sigma' \quad , \quad \gamma = \tau'. \tag{6.15}$$

Then, using (6.13), we can integrate (6.10) to obtain

$$\underset{\sim}{x} = \underset{\sim}{k} + \underset{\sim}{R} \, [\underset{\sim}{y}(X^{\alpha}) - \underset{\sim}{k}] + \sigma \underset{\sim}{e} , \tag{6.16}$$

where $\underset{\sim}{y} \, (X^{\alpha})$ represents arbitrary functions arising in the integration and

$$\underset{\sim}{R} = \exp \tau \underset{\sim}{\Omega} , \tag{6.17}$$

wherein $\underset{\sim}{\Omega}$ is related to $\underset{\sim}{e}$ by (3.4). Now differentiation of (6.16) gives

$$\begin{cases} \underset{\sim}{x}_{,\alpha} = \underset{\sim}{R}\,\underset{\sim}{y}_{,\alpha}\,, \\ \underset{\sim}{x}_{,3} = \underset{\sim}{R}\,\underset{\sim}{p}\,, \\ \underset{\sim}{p} = \gamma\,\underset{\sim}{e} \wedge (\underset{\sim}{y} - \underset{\sim}{k}) + \alpha\,\underset{\sim}{e}, \end{cases} \tag{6.18}$$

much as for the semi-inverse solutions, except that α and γ might here depend on X^3. Thus, as before,

$$W(\underset{\sim}{x}_{,\alpha}\ ;\ \underset{\sim}{x}_{,3}\ ;\ X^{\beta}) = W(\underset{\sim}{y}_{,\alpha}\ ;\ \underset{\sim}{p}\ ;\ X^{\beta}).$$

If we now add the conditions that both (6.9) and the equilibrium equations be satisfied, we have

$$\begin{aligned} 0 &= \left(\frac{\partial W}{\partial \underset{\sim}{x}_{,K}}\right)_{,K} \\ &= \left(\frac{\partial W}{\partial \underset{\sim}{x}_{,\alpha}}\right)_{,\alpha} + \underset{\sim}{R}\left[\gamma\,\underset{\sim}{e} \wedge \frac{\partial W}{\partial \underset{\sim}{p}} + \left(\frac{\partial W}{\partial \underset{\sim}{p}}\right)_{,3}\right] \\ &= \left(\frac{\partial W}{\partial \underset{\sim}{x}_{,\alpha}}\right)_{,\alpha} + \gamma\,\underset{\sim}{e} \wedge \frac{\partial W}{\partial \underset{\sim}{x}_{,3}} + \underset{\sim}{R}\left(\frac{\partial W}{\partial \underset{\sim}{p}}\right)_{,3} \\ &= \underset{\sim}{R}\left(\frac{\partial W}{\partial \underset{\sim}{p}}\right)_{,3}. \end{aligned}$$

Thus

$$\left(\frac{\partial W}{\partial \underset{\sim}{p}}\right)_{,3} = \frac{\partial^2 W}{\partial \underset{\sim}{p}\,\partial \underset{\sim}{p}}\,\underset{\sim}{p}_{,3} = \underset{\sim}{0}. \tag{6.19}$$

With (6.8), it follows that (6.19) can hold only if $\underset{\sim}{p}$ is independent of X^3 and using (6.18), this implies that α and γ must be constants. Thus the semi-inverse solutions can be characterized as extremals of g, in the sense described above. The analysis suggests that g slowly varies in the neighborhood of a semi-inverse solution, if we respect the constraint on faces and moments. Perhaps, it is this which enhances their ability to represent a broader set of solutions than might at first seem reasonable. For solutions such that g is not close to an extremal value, our semi-inverse solutions must give way to some other type of representative solutions. Perhaps, this is why the St.-Venant solutions for flexure are cut from different cloth.

Acknowledgment

This paper summarizes lectures given while the author visited the Mathematics Department at Heriot-Watt University. He much appreciated the warm hospitality and lively discussions which attended this visit.

References

1 E. Sternberg and J. K. Knowles, Minimum Energy Characterization of St.-Venant's solution to the relaxed Saint-Venant Problem, Arch. Rational Mech. Anal. (1966) 21, 89-107.

2 J. L. Ericksen, Special Topics in Elastostatics, to appear in Advances Applied Mech.

3 J. L. Ericksen, Transformation Theory for some Continuum Theories, (to appear in Iran J. Sci. Tech.).

4 A. E. H. Love, A Treatise on the Mathematical Theory of Elasticity, 4th ed., Cambridge University Press (1927).

5 E. Noether, Invariante Variationsprobleme, Nachr. Akad. Wiss. Göttingen-Math.-Phys. Klasse 2, (1918), 235-257.

6 R. A. Toupin, Saint-Venant's Principle. Arch. Rational Mech. Anal. (1955), 18, 83-96.

Professor J. L. Ericksen, Department of Mechanics and Materials Science,
The Johns Hopkins University, Baltimore, Md. 21218, U.S.A.

J M BALL

Constitutive inequalities and existence theorems in nonlinear elastostatics

In these notes I shall describe an approach to the problem of proving the existence of equilibrium solutions in nonlinear elasticity. This problem confronts the analyst with a difficulty typical of nonlinear continuum mechanics, that of choosing hypotheses on the constitutive equations which are both physically reasonable and ensure the existence of solutions with the desired degree of smoothness. The study of constitutive equations in elasticity from the point of view of existence of solutions leads one to consider new constitutive inequalities, and gives insight which it would be hard to gain by other methods.

One-dimensional elasticity

To focus attention on the rôle played by constitutive inequalities in questions of existence, we begin by discussing one-dimensional nonlinear elasticity, which is markedly simpler to treat analytically than the three-dimensional case.

Consider a one-dimensional elastic body (you can think of it as a thin bar) which occupies the unit interval $0 < X < 1$ in a reference configuration (see Figure 1a). In a typical deformed configuration (see Figure 1b) the particle P with position X moves to the point P' having coordinate $x(X)$ with respect to some fixed origin.

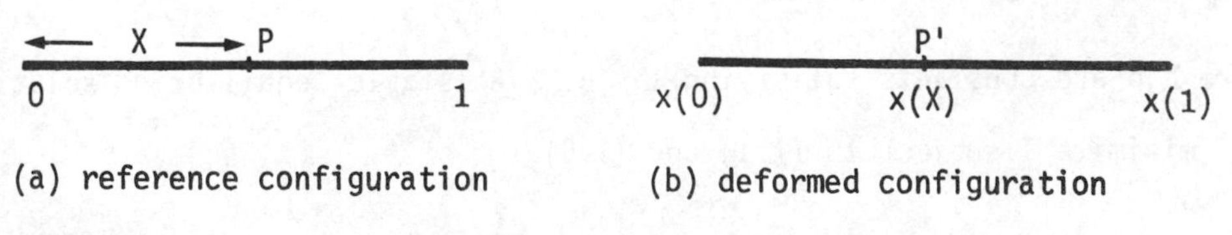

Figure 1

Because we are considering elastostatics, x is assumed to be independent of time. We are interested in deformations $x(X)$ satisfying the <u>invertibility condition</u>

$$x'(X) > 0, \quad 0 < X < 1, \tag{1.1}$$

where the prime denotes differentiation with respect to X. Condition (1.1) prevents interpenetration of matter. We assume that the mechanical behaviour of the material is characterized by a stored-energy function $W(X,x')$, in terms of which the total stored-energy is

$$E(x) = \int_0^1 W(X,x'(X))dX. \tag{1.2}$$

Here we have ignored thermal effects. If the body force is conservative with a continuous potential $\psi(x)$ then the equilibrium equation is the Euler-Lagrange equation for the functional

$$I(x) = E(x) + \int_0^1 \psi(x(X))dX, \tag{1.3}$$

namely

$$\frac{\partial\psi}{\partial x} - \frac{d}{dX}\left(\frac{\partial W}{\partial x'}\right) = 0. \tag{1.4}$$

In a <u>displacement</u> boundary-value problem (BVP) we have to solve (1.4) subject to the invertibility condition (1.1) and the boundary conditions

$$x(0) = a, \quad x(1) = b \tag{1.5}$$

where a,b are constants satisfying $a < b$. A 'stable' equilibrium solution will minimize I subject to (1.1) and (1.5).

Whether or not a minimizer exists depends on the form of W. Now it is commonly observed that a rod lengthens when subjected to a tensile force, that is, stress increases with strain. The one-dimensional stress is simply $\sigma = \frac{\partial W}{\partial x'}$, which means that W should be convex in x'. We examine the consequences of assuming this to be so. A typical stress-strain curve for rubber is sketched in Figure 2a, with the corresponding stored-energy function W, assumed independent of X, shown in Figure 2b. The reference configuration is assumed stress-free, so that $\sigma(1) = 0$, and we have taken $W(1) = 0$ without loss of generality. Note that W and σ become large in magnitude for both large and small values of x'; this reflects the fact that a large force is

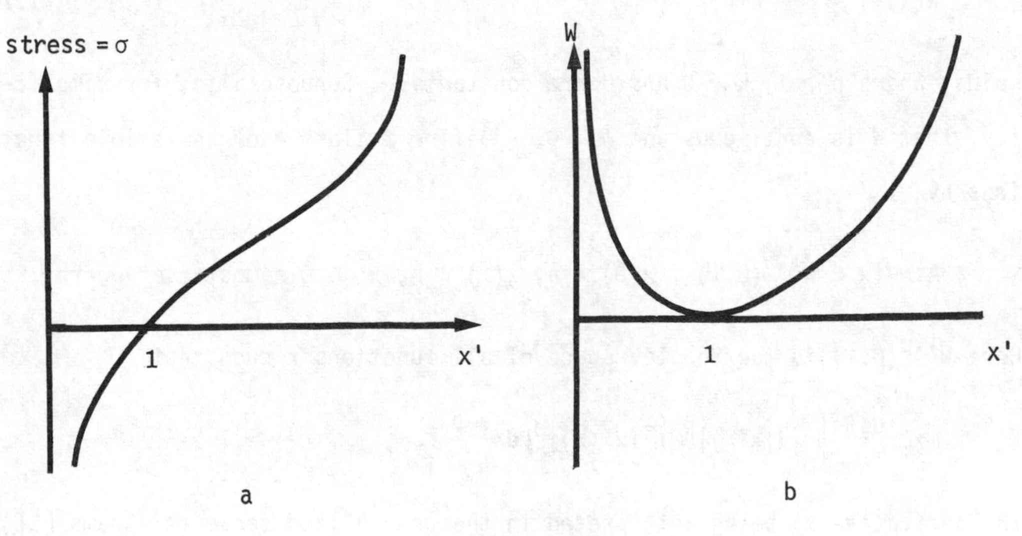

Figure 2

required to effect a large extension or compression. If we take the (somewhat unrealistic) view that any value of $x' > 0$ is possible, then it seems reasonable to assume that for each $X \in (0,1)$

$$W(X,x') \longrightarrow \infty \quad \text{as} \quad x' \longrightarrow 0, \infty. \tag{1.6}$$

Hopefully the first of these requirements will ensure that (1.1) holds.

The convexity of W implies that equation (1.4) is <u>elliptic</u>. Let us sketch a proof of the existence of a minimizer using the 'direct method' of the calculus of variations. We augment (1.6) by assuming that the growth condition

$$W(X,v) \geqslant C + k|v|^p \tag{1.7}$$

holds, where $p > 1$, $k > 0$ and C are constants. Suppose also, for simplicity, that W is continuous and $\psi \geqslant 0$. Define a class A of admissible functions by

$$A = \{x \in W^{1,p}(0,1) : x(0) = a, \ x(1) = b, \ x' \geqslant 0 \text{ almost everywhere}\}.$$

Here $W^{1,p}(0,1)$ is the Sobolev space of all functions x such that

$$\|x\| \stackrel{\text{def}}{=} \left(\int_0^1 [|x(X)|^p + |x'(X)|^p] \, dX\right)^{1/p} < \infty,$$

the derivative x' being interpreted in the generalized sense (cf Adams [1]). $W^{1,p}(0,1)$ is a reflexive Banach space. Let $\{x_n\} \subset A$ be a minimizing sequence for I. Using (1.7) and the Poincaré inequality we find that $\|x_n\|$ is bounded, so that a subsequence $\{x_\mu\}$ exists converging weakly in $W^{1,p}(0,1)$ to a function x_0. Since A is weakly closed it follows that $x_0 \in A$. Since W is convex in x' and ψ is continuous we deduce by standard arguments that I is sequentially weakly lower semicontinuous (swlsc), i.e.,

$$I(x_o) \leqslant \liminf_{\mu \longrightarrow \infty} I(x_\mu).$$

Hence x_o is a minimizer for I in A. By (1.6), $x_o' > 0$ almost everywhere. To show that x_o is smooth with $x_o' \geqslant \delta > 0$ in (0,1) is a tricky piece of regularity theory, requiring further hypotheses on W and ψ. The reader is referred to Antman [2,3,4] for details of how this can be done.

The above argument relies crucially on the convexity of W; indeed it has been known since the work of Tonelli that under suitable growth conditions convexity is essentially a necessary and sufficient condition for I to be swlsc. What is more, a simple argument shows that the convexity of W is a necessary condition for the existence of C^1 minimizers for all displacement BVPs of the above type. Consider the case $\psi \equiv 0$, and assume for simplicity that W is C^2 and does not depend explicitly on X. Take a = 0, b > 0 and suppose that $x_o \in C^1([0,1])$ minimizes E(x) among all functions $x \in C^1([0,1])$ satisfying x(0) = 0, x(1) = b, x'(X) > 0 for all $X \in [0,1]$. For any $y \in \mathcal{D}(0,1)$, the set of infinitely differentiable functions with compact support in (0,1), we obtain

$$\frac{d^2}{d\varepsilon^2} E(x_0+\varepsilon y)\Big|_{\varepsilon=0} = \int_0^1 \frac{\partial^2 W}{\partial x'^2}(x_0'(X))y'(X)^2 dx \geqslant 0.$$

Hence, $\frac{\partial^2 W}{\partial x'^2}(x_0(X)) \geqslant 0$ for all $X \in [0,1]$. But by the mean value theorem there exists $\bar{X} \in (0,1)$ with $x_o'(\bar{X}) = b$. Hence, $\frac{\partial^2 W}{\partial x'^2}(b) \geqslant 0$ and the arbitrariness of b implies that W is convex.

Note that if W is not convex then there may exist minimizers for I that are not C^1. The assumption of convexity of W can be criticized on the grounds that it is not generic, i.e., convexity is not preserved under small perturbations. Indeed, nonconvex W are of physical interest, since they may be associated with materials that undergo phase transitions (Ericksen [12-14]). Nevertheless convexity plays an important rôle even for nonconvex W, since in this case a 'relaxation theorem' of Ekeland and Témam [10] shows that, under certain conditions, from any minimizing sequence for I a subsequence may be extracted converging weakly in a Sobolev space to a minimizer for the functional

$$\bar{I}(x) = \int_0^1 \bar{W}(x'(X))dX + \int_0^1 \psi(x(X))dX,$$

where $\bar{W}$ denotes the lower convex envelope of W (see Figure 3).

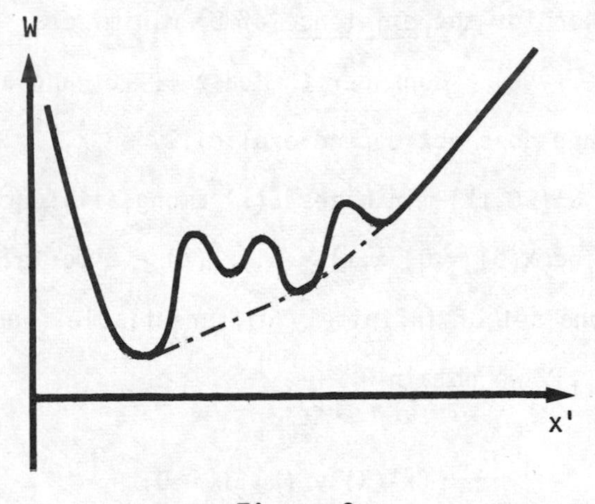

Figure 3

We will not consider this interesting point of view further here.

Three-dimensional elasticity

Consider an elastic body which in a reference configuration occupies the bounded open set $\Omega \subset \mathbb{R}^3$. In a typical deformed configuration the particle P with position vector $\underset{\sim}{X}$ moves to the point P' having position vector $\underset{\sim}{x}(\underset{\sim}{X})$ with respect to fixed Cartesian axes (Figure 4).

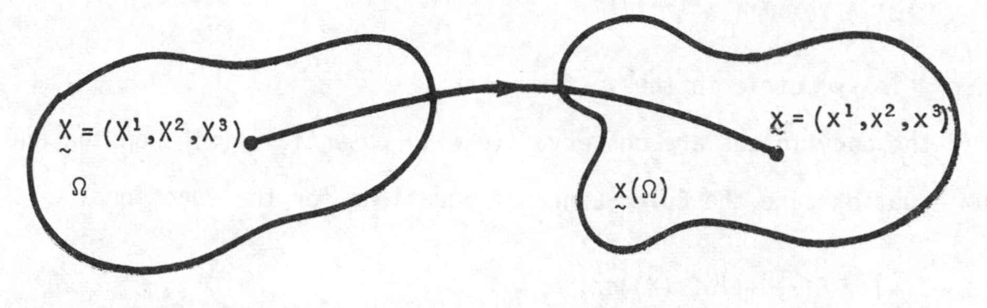

Reference configuration   Deformed configuration

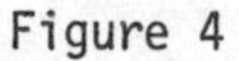

Figure 4

The <u>deformation gradient</u> F is defined by

$$F = \nabla \underset{\sim}{x} \; ; \quad F^i_\alpha = x^i_{,\alpha}.$$

We are interested only in deformations which are orientation preserving and globally invertible. In particular we require the <u>local invertibility condition</u>

$$\det F > 0 \quad \text{for all } \underset{\sim}{X} \in \Omega \tag{1.8}$$

to hold. By the polar decomposition theorem $F=RU$ for some proper orthogonal matrix R and some positive definite symmetric matrix U. The positive eigenvalues v_i $(i = 1,2,3)$ of U are called the <u>principal stretches</u> of the deformation.

The mechanical properties of the material are characterized by a stored-energy function $W(\underset{\sim}{X},F)$ in terms of which the total stored-energy is

$$E(\underset{\sim}{x}) = \int_\Omega W(\underset{\sim}{X}, \nabla \underset{\sim}{x}(\underset{\sim}{X}))d\underset{\sim}{X}.$$

If the material is isotropic then W has the form

$$W(\underset{\sim}{X},F) = \Phi(\underset{\sim}{X},v_1,v_2,v_3), \tag{1.9}$$

where Φ is symmetric in the v_i for each $\underset{\sim}{X}$.

If the body forces are conservative with potential $\psi(\underset{\sim}{x})$ then the equilibrium equations are the Euler-Lagrange equations for the functional

$$I(\underset{\sim}{x}) = E(\underset{\sim}{x}) + \int_{\Omega} \psi(\underset{\sim}{x}(\underset{\sim}{X}))d\underset{\sim}{X},$$

namely

$$\frac{\partial\psi}{\partial x^i} - \frac{\partial}{\partial X^\alpha}\frac{\partial W}{\partial F^i_\alpha} = 0. \tag{1.10}$$

(Repeated indices are summed from 1 to 3). Let us consider a mixed BVP in which $\underset{\sim}{x}$ is prescribed on a portion $\partial\Omega_1$ of the boundary $\partial\Omega$ of Ω, so that

$$\underset{\sim}{x}(\underset{\sim}{X}) = \bar{\underset{\sim}{x}}(\underset{\sim}{X}) \quad \text{for } \underset{\sim}{X} \in \partial\Omega_1, \tag{1.11}$$

and in which the remainder of the boundary of the body is free of applied surface forces. We seek a minimizer of I subject to (1.8) and (1.11). We do not need to worry about the zero traction condition on $\partial\Omega\backslash\partial\Omega_1$, since this is a natural boundary condition.

What hypotheses should we make on W? Bearing in mind our experience with the one-dimensional case, the natural first try is to assume that $W(\underset{\sim}{X},\cdot)$ is convex; if we make this assumption then it is not hard to generalize our previous analysis from one to three dimensions and to prove various existence theorems. The only trouble is that convexity of $W(\underset{\sim}{X},\cdot)$ is completely unrealistic. There are a number of ways in which this can be seen. Suppose for simplicity that W does not depend explicitly on $\underset{\sim}{X}$, and that $\psi \equiv 0$. If

W(F) were convex and C^1 on some open convex subset S of the space $M^{3\times 3}$ of 3×3 matrices, then $E(\underset{\sim}{x})$ would be a Gâteaux differentiable convex function of $\underset{\sim}{x}$ on the relatively open convex set $K = \{\underset{\sim}{x} \in C^1(\bar{\Omega})$: $\underset{\sim}{x}$ satisfies (1.11) and $\nabla \underset{\sim}{x}(\underset{\sim}{X}) \in S$ for all $\underset{\sim}{X} \in \bar{\Omega}\}$. A result on convex functions (cf Ekeland and Témam [10 Prop. 5.3]) then implies that the set of equilibrium solutions

$$\{\underset{\sim}{x} \in K : E'(\underset{\sim}{x}) = 0\}$$

is convex and consists of absolute minimizers for E on K. In particular, W strictly convex implies the uniqueness of equilibrium solutions in K. Thus convexity rules out the multiple solutions and instabilities commonly observed in buckling. (For examples of nonuniqueness in elastostatics see Rivlin [29,31], John [19,20], Wang and Truesdell [41]). Similar considerations apply in other function spaces. Convexity of W also conflicts with the natural requirement that W be frame-indifferent i.e.,

$$W(\underset{\sim}{X},QF) = W(\underset{\sim}{X},F) \quad \text{for all proper orthogonal matrices } Q. \tag{1.12}$$

For details see Coleman and Noll [8], Truesdell and Noll [38]. Finally, the inappropriateness of convexity as a constitutive assumption can be seen from the following experiment. Take a homogeneous, isotropic, rubber sheet and subject it to a homogeneous deformation in which $F = \text{diag}(v_1, v_2, 1)$. Consider the stored-energy function W as a function of v_1 and v_2. As rubber is almost incompressible it takes a lot of energy to produce a deformation in which $v_1 v_2$ differs greatly from 1. Thus the contours of equal energy are banana-shaped as in Figure 5. This is not consistent with convexity of W. As an illustration, deformations in which $v_1 = 4$, $v_2 = \frac{1}{4}$ or $v_1 = \frac{1}{4}$, $v_2 = 4$ would be easy to produce with one's bare hands, but the case $v_1 = v_2 = \frac{17}{8}$, if possible to achieve at all, would require much more energy. Convexity of W would imply that $W\left(\frac{17}{8},\frac{17}{8}\right) \leqslant W(1,4) = W(4,1)$.

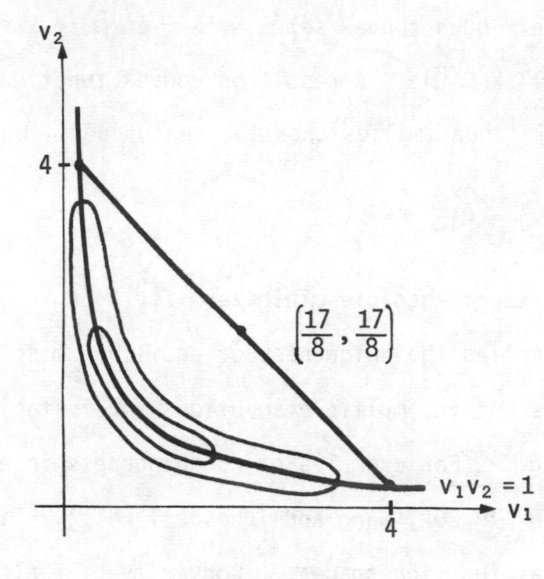

Figure 5

Having disposed of convexity we must search for a suitable substitute. There is a large literature on various alternative convexity hypotheses proposed more or less on ad hoc grounds (see Wang and Truesdell [41] for a summary), and there is no general agreement as to which is the most appropriate†. Whereas in one dimension the intuitive significance of stress increasing with strain is clear, in three dimensions it is not obvious precisely which combinations of surface forces will effect, say, an increase in volume of a unit cube made of <u>any</u> elastic material, and most of the ad hoc inequalities are based on plausibility arguments of this type. An example is the Coleman-Noll condition (see Coleman and Noll [8], Wang and Truesdell [41]). In the case of an isotropic material it implies that $\Phi(\underset{\sim}{X},v_1,v_2,v_3)$ given by (1.9) is convex in

† The general problem of finding suitable constitutive inequalities for nonlinear elasticity was originally posed by Truesdell [37].

the v_i. For rubber this is ruled out by Figure 5 (see also Rivlin [30], Ogden [24]), and for this reason we do not adopt the Coleman-Noll condition.

An older and better motivated inequality is the Legendre-Hadamard or ellipticity condition, which in the case of a smooth stored-energy function requires that

$$\frac{\partial^2 W(\underset{\sim}{X},F)}{\partial F^i_\alpha \partial F^j_\beta}\lambda^i\lambda^j\mu_\alpha\mu_\beta \geqslant 0 \qquad \text{for all } \underset{\sim}{\lambda},\ \underset{\sim}{\mu} \in \mathbb{R}^3. \tag{1.13}$$

Hadamard's theorem [17,18] asserts that (1.13) holds in Ω for any C^1 minimizer of our mixed BVP. If we suppose that (1.13) holds for all $\underset{\sim}{X}$ and F then, as the name suggests, (1.10) becomes an elliptic system. Also $W(\underset{\sim}{X},\cdot)$ need not be convex, so that multiple equilibrium solutions may be possible. Unfortunately, while there is a well developed existence and regularity theory for linear elliptic systems, no such theory seems to be known for non-linear elliptic systems. However it can be shown (Theorem 2.1) that if $\underset{\sim}{x}$ is a C^1 minimizer for our BVP then another inequality implying (1.13) holds, namely the quasiconvexity condition of Morrey. This states that

$$\int_D W(\underset{\sim}{X},F+\nabla\underset{\sim}{\zeta}(\underset{\sim}{Y}))d\underset{\sim}{Y} \geqslant \int_D W(\underset{\sim}{X},F)d\underset{\sim}{Y} = W(\underset{\sim}{X},F) \times \text{volume of } D \tag{1.14}$$

for all $\underset{\sim}{X} \in \Omega$, $F = \nabla\underset{\sim}{x}(\underset{\sim}{X})$, all bounded open subsets D of $\mathbb{R}^3$ and all $\underset{\sim}{\zeta} \in \mathcal{D}(D)$. If we turn (1.14) into a constitutive hypothesis by requiring it to hold for all $\underset{\sim}{X} \in \Omega$, $F \in M^{3\times 3}$ then under certain other growth and continuity conditions $I(\underset{\sim}{x})$ is swlsc and it is possible to prove the existence of equilibrium solutions. This follows from the work of Morrey [22].

The quasiconvexity condition has an interesting physical interpretation that follows immediately from (1.14): For any homogeneous body made from the material found at any point of Ω, and for any displacement BVP with zero body force for such a body that admits as a possible deformation a homogeneous strain, this homogeneous strain must be an absolute minimizer for the total energy. Note that this interpretation would not be in accord with experience if inhomogeneous bodies or mixed BVPs were allowed, as we would expect certain buckled states to have lower energy than the homogeneous strain. Consider for example a displacement BVP for a solid cube consisting of a steel bar imbedded in a rubber matrix. If the homogeneous strain is a compression in the direction of the bar axis then buckling can occur. See Figure 6.

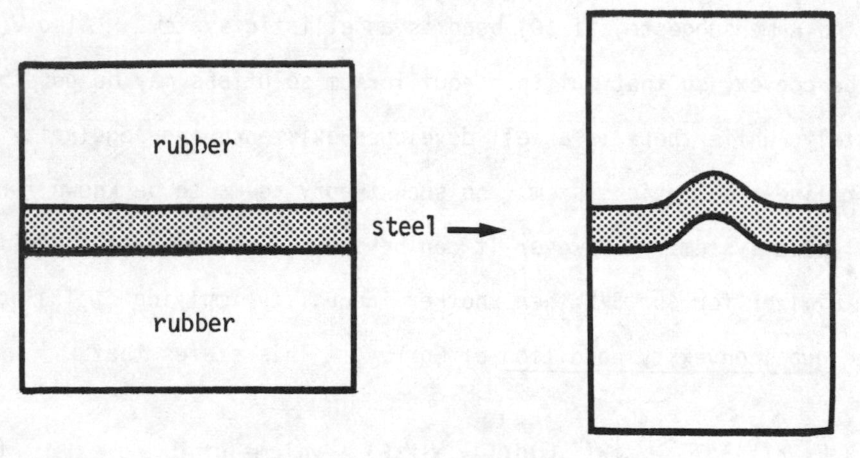

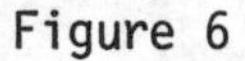

Figure 6

Morrey's existence theorem is very interesting, but one encounters difficulties when applying it to elasticity because of the very strong continuity and growth assumptions used for the proof. In particular $W(\underset{\sim}{X},F)$ is supposed to be defined and continuous for all F. This rules out any singularities of W such as the natural condition

$$W(\underset{\sim}{X},F) \longrightarrow \infty \quad \text{as} \quad \det F \longrightarrow 0. \tag{1.15}$$

(Note that (1.15) is not consistent with convexity of $W(\underset{\sim}{X},\cdot)$). Furthermore the one-dimensional case suggests that we will need to seek a minimum for I on a set of the form

$$A = \{\underset{\sim}{x} \in W^{1,p}(\Omega) : \underset{\sim}{x} = \bar{\underset{\sim}{x}} \text{ on } \partial\Omega_1, \det \nabla\underset{\sim}{x} \geqslant 0 \text{ almost everywhere}\}$$

for some $p > 1$, and it is not obvious that such a set will be sequentially weakly closed in $W^{1,p}(\Omega)$. Similar problems arise in the important case of <u>incompressible</u> elasticity, where we have to satisfy the constraint

$$\det F = 1. \tag{1.16}$$

A key to the resolution of these difficulties lies in the concept of a <u>null Lagrangian</u>. A continuous function $\phi: M^{3\times3} \longrightarrow \mathbb{R}$ is a null Lagrangian if the Euler-Lagrange equations for the functional $\int_\Omega \phi(\nabla\underset{\sim}{x}(\underset{\sim}{X}))d\underset{\sim}{X}$ reduce to $0 = 0$; i.e., they are identically satisfied for every $\underset{\sim}{x} \in C^1(\bar{\Omega})$. Equivalently

$$\int_\Omega \phi(\nabla\underset{\sim}{x}(\underset{\sim}{X}) + \nabla\underset{\sim}{\zeta}(\underset{\sim}{X}))d\underset{\sim}{X} = \int_\Omega \phi(\nabla\underset{\sim}{x}(\underset{\sim}{X}))d\underset{\sim}{X} \tag{1.17}$$

for all $\underset{\sim}{x} \in C^1(\bar{\Omega})$, $\underset{\sim}{\zeta} \in \mathcal{D}(\Omega)$. Clearly if ϕ is a null Lagrangian then (1.10) is invariant under the transformation $W \longmapsto W + \phi$. In particular, all displacement BVP's for W and $W + \phi$ have the same solutions. We will show in Section 3 that ϕ is a null Lagrangian if and only if

$$\phi(F) = A + A_i^\alpha F_\alpha^i + B_i^\alpha (\operatorname{adj} F)_\alpha^i + C \det F, \tag{1.18}$$

where A, A_i^α, B_i^α, C are constants, and where $\operatorname{adj} F$ denotes the transpose of the matrix of cofactors of F. For our purposes the importance of null Lagrangians lies in various versions of the following result: if $p > 3$ and $|\phi(F)| \leqslant C(1 + |F|^p)$ for some constant $C > 0$ and all $F \in M^{3\times3}$, then the map

$\phi(\nabla \underset{\sim}{x}(\cdot)) : W^{1,p}(\Omega) \longrightarrow L^1(\Omega)$ is sequentially weakly continuous (i.e., $\underset{\sim}{x}_n \longrightarrow \underset{\sim}{x}$ in $W^{1,p}(\Omega)$ implies $\phi(\nabla \underset{\sim}{x}_n(\cdot)) \longrightarrow \phi(\nabla \underset{\sim}{x}(\cdot))$ if and only if ϕ is a null Lagrangian. Related weak continuity results have been studied in so far unpublished work of F. Murat and L. Tartar (1974) and L. Tartar (1976).

A simple constitutive hypothesis which is invariant under the transformation $W \longmapsto W + \phi$ for every null Lagrangian ϕ is the following: there exists a function $g : \Omega \times M^{3\times 3} \times M^{3\times 3} \times (0,\infty) \longrightarrow \mathbb{R}$ with $g(\underset{\sim}{X},\cdot,\cdot,\cdot)$ convex for each $\underset{\sim}{X} \in \Omega$ such that

$$W(\underset{\sim}{X},F) = g(\underset{\sim}{X},F,\mathrm{adj}\,F,\det F)$$

for all $\underset{\sim}{X} \in \Omega$ and all $F \in M^{3\times 3}$ with $\det F > 0$. We call such functions W polyconvex. It is easily shown that polyconvex functions are quasiconvex, but a polyconvex function need not be convex in F, as the example $W(F) = \det F$ shows. The results on sequential weak continuity mentioned above imply that if W is polyconvex, then under suitable growth and continuity assumptions $I(\underset{\sim}{x})$ is swlsc on $W^{1,p}(\Omega)$. Also, sets of the form A above are sequentially weakly closed. As in one dimension this leads quickly to a proof of the existence of minimizers.

We now turn to the question of what are appropriate growth conditions for W. In the case of compressible elasticity we assume that (1.15) holds. Some condition analogous to (1.7) is also required to restrict the behaviour of W for large $|F|$. Consider a cube of side $\frac{1}{\lambda}$ made from the material found at $\underset{\sim}{X}$. For fixed $F \in M^{3\times 3}$ with $\det F > 0$, imagine deforming the cube by a homogeneous strain with deformation gradient λF. The shape and size of the deformed cube is independent of λ. A plausible requirement is that as $\lambda \longrightarrow \infty$ the total energy of the deformation becomes unbounded, that is

$$\frac{1}{\lambda^3} W(\underset{\sim}{X},\lambda F) \longrightarrow \infty \quad \text{as} \quad \lambda \longrightarrow \infty. \tag{1.19}$$

The stronger condition

$$\frac{W(\underset{\sim}{X},F)}{|F|^3} \longrightarrow \infty \quad \text{as} \quad |F| \longrightarrow \infty \tag{1.20}$$

says that a line segment of positive length cannot be produced from an infinitesimal cube using a finite amount of energy. A sufficient condition for (1.20) to hold is obviously that

$$W(\underset{\sim}{X},F) \geqslant a(\underset{\sim}{X}) + k\,|F|^{3+\varepsilon}, \tag{1.21}$$

for some function $a(\cdot)$, and constants $k > 0$, $\varepsilon > 0$. If (1.21) holds then for certain problems a minimizer $\underset{\sim}{x}$ exists in the space $W^{1,3+\varepsilon}(\Omega)$. The Sobolev imbedding theorem then implies that $\underset{\sim}{x}$ is continuous. This fits in nicely with our motivation for (1.19) and (1.20), since these conditions should prevent holes being formed in the body. In practice we will use a variety of growth conditions; in some cases one can prove existence under conditions that do not imply (1.20).

In the case of incompressible elasticity we assume that $W(\underset{\sim}{X},F)$ is defined for all F with $\det F = 1$. An analogue of (1.20) is that

$$\frac{W(\underset{\sim}{X},F)}{|F|^3} \longrightarrow \infty \quad \text{as} \quad |F| \longrightarrow \infty \quad \text{with } \det F = 1. \tag{1.22}$$

Conditions like (1.20) and (1.22) are especially important for pressure BVP's. For example it is easy to verify that the total energy functional for a spherical shell of Neo-Hookean material ($W(F) = \alpha(\mathrm{tr}(FF^T)-3)$, $\alpha > 0$), under constant internal and external pressures $p > 0$ and zero respectively, is not bounded below, so that no absolute minimizer exists.

As has been indicated above, the existence theorems proved in these notes are of a global type and apply to problems with multiple equilibrium solutions. It is also possible to prove local results, in which the existence and uniqueness of small solutions to BVP's with small body forces and boundary data are established via the inverse function theorem. This has been done by Stoppelli [35] and van Buren [39] (see also Truesdell and Noll [38], Wang and Truesdell [41]). The material response is assumed to be such that existence, uniqueness and regularity theorems hold for the equilibrium equations linearized about the zero data solution. Although these results are limited in scope, they do only need assumptions about the material response for deformation gradients close to those of the zero data solution.

The plan of the remainder of these notes is as follows. In section 2 we discuss in detail the quasiconvexity, ellipticity and polyconvexity conditions. In section 3 we prove the results concerning sequential weak continuity and null Lagrangians. The main existence theorems are given in section 4. In section 5 we comment on the problem of proving that minimizers are smooth and indicate also how the existence theorems in section 4 may be extended to apply to stored-energy functions of 'slow growth' (this involves the use of 'distributional' determinants). In section 6 the existence theorems are applied to various models of rubber. Finally, in section 7 we prove the existence of minimizers for semi-inverse problems of the type recently introduced by Ericksen.

Much of the material presented here appeared first in [A]. Exceptions are the conditions (1.19), (1.20), Theorem 2.6 and the whole of section 7. However the presentation is different (and I hope more readable), and I have tried throughout to use the simplest possible technical assumptions. For example, the existence theorems are proved in Sobolev spaces, rather than in the more general Orlicz-Sobolev spaces used in places in [A].

2 CONSTITUTIVE INEQUALITIES

Let Ω be a bounded open subset of $\mathbb{R}^3$. Let $M^{3\times3}$ have the induced norm of $\mathbb{R}^9$, and let U be the open subset of $M^{3\times3}$ consisting of F with $\det F > 0$. Consider a continuous stored-energy function $W : \Omega \times U \longrightarrow \mathbb{R}$.

<u>Definition</u> (cf Morrey [22,23]) W is <u>quasiconvex at a point</u> $(\underset{\sim}{X},F) \in \Omega \times U$ if and only if

$$\int_D W(\underset{\sim}{X},F+\nabla\underset{\sim}{\zeta}(\underset{\sim}{Y}))d\underset{\sim}{Y} \geqslant W(\underset{\sim}{X},F) \times \text{volume of } D \tag{2.1}$$

for every bounded open subset $D \subset \mathbb{R}^3$ and for every $\underset{\sim}{\zeta} \in \mathfrak{D}(D)$ satisfying $F + \nabla\underset{\sim}{\zeta}(\underset{\sim}{Y}) \in U$ for all $\underset{\sim}{Y} \in \Omega$. W is <u>quasiconvex</u> if it is quasiconvex at every $(\underset{\sim}{X},F) \in \Omega \times U$.

If W is quasiconvex at $(\underset{\sim}{X},F)$ then an approximation argument shows that (2.1) holds for any $\underset{\sim}{\zeta} \in C^1(\bar{D})$ with $\underset{\sim}{\zeta} = \underset{\sim}{0}$ on ∂D and such that $F + \nabla\underset{\sim}{\zeta}(\underset{\sim}{Y}) \in D$ for all $\underset{\sim}{Y} \in \bar{D}$.

Let

$$E(\underset{\sim}{x}) = \int_\Omega W(\underset{\sim}{X},\nabla\underset{\sim}{x}(\underset{\sim}{X}))d\underset{\sim}{X},$$

and let $A = \{\underset{\sim}{x} \in C^1(\bar{\Omega}) : \nabla\underset{\sim}{x}(\underset{\sim}{X}) \in U \text{ for all } \underset{\sim}{X} \in \bar{\Omega}\}$. For each $\underset{\sim}{x} \in A$, $\nabla\underset{\sim}{x}(\bar{\Omega})$ is a compact subset of U. Since W is continuous this implies that $E: A \longrightarrow \mathbb{R}$.

The following extension of Hadamard's theorem motivates the quasiconvexity condition by showing that the condition is satisfied at every point of a minimizer for a displacement BVP with zero body force. A similar result is stated by Silverman [34]; see also Busemann and Shephard [7].

Theorem 2.1 Let $\bar{x} \in A$ satisfy

$$E(x) \geqslant E(\bar{x})$$

for all $x \in A$ with $x - \bar{x} \in \mathcal{D}(\Omega)$ and $\|x-\bar{x}\|_{C(\bar{\Omega})} \overset{\text{def}}{=} \sup_{X \in \Omega} |x(X) - \bar{x}(X)|$ sufficiently small. Then W is quasiconvex at every point $(X, \nabla\bar{x}(X))$, $X \in \Omega$.

Proof. Let D be a bounded open subset of $\mathbb{R}^3$, let $X_o \in \Omega$, and let $\zeta \in \mathcal{D}(D)$ satisfy $\nabla\bar{x}(X_o) + \nabla\zeta(Y) \in U$ for all $Y \in D$. For $\varepsilon > o$ define $x_\varepsilon : \Omega \longrightarrow \mathbb{R}^3$ by

$$x_\varepsilon(X) = \begin{cases} \bar{x}(X) + \varepsilon\zeta\left(\dfrac{X - X_o}{\varepsilon}\right) & \text{if } \dfrac{X-X_o}{\varepsilon} \in D \\ \bar{x}(X) & \text{otherwise.} \end{cases}$$

For small enough ε the set $X_o + \varepsilon D$ is contained in Ω, so that $x_\varepsilon - \bar{x} \in \mathcal{D}(\Omega)$. Also

$$\nabla x_\varepsilon(X) = \begin{cases} \nabla\bar{x}(X) + \nabla\zeta\left(\dfrac{X - X_o}{\varepsilon}\right) & \text{if } \dfrac{X-X_o}{\varepsilon} \in D \\ \nabla\bar{x}(X) & \text{otherwise.} \end{cases}$$

The continuity assumptions therefore imply that $x_\varepsilon \in A$. Since $\|x_\varepsilon - \bar{x}\|_{C(\bar{\Omega})} \longrightarrow 0$ as $\varepsilon \longrightarrow 0$ it follows that $E(x_\varepsilon) \geqslant E(\bar{x})$ for small enough ε. Making the change of variables $Y = \dfrac{X - X_o}{\varepsilon}$ and dividing by ε^3 we obtain

$$\int_D W(X_o + \varepsilon Y, \nabla\bar{x}(X_o + \varepsilon Y) + \nabla\zeta(Y))\,dY \geqslant \int_D W(X_o + \varepsilon Y, \nabla\bar{x}(X_o + \varepsilon Y))\,dY.$$

Now let $\varepsilon \longrightarrow 0$ to get the result. □

It follows from the theorem that if the quasiconvexity condition (2.1) holds for <u>one</u> bounded open subset $D \subset \mathbb{R}^3$ and all $(\underset{\sim}{X},F) \in \Omega \times U$, then it holds for <u>all</u> such subsets. To see this one simply applies the theorem with $\Omega = D$ and $\bar{\underset{\sim}{x}}(\underset{\sim}{X}) = F\underset{\sim}{X}$.

Note that in <u>one dimension</u> quasiconvexity is equivalent to convexity. To be precise let $W : (0,1) \times (0,\infty) \longrightarrow \mathbb{R}$ satisfy

$$\int_a^b W(X,F+\zeta'(Y))dY \geqslant W(X,F)(b-a) \tag{2.2}$$

for every bounded open interval (a,b), every $\zeta \in \mathfrak{D}(a,b)$ satisfying $F + \zeta'(Y) > 0$ for all $Y \in (a,b)$, and every $X \in (0,1)$, $F > 0$. Let $\varepsilon > 0$, $H > 0$, $\lambda \in [0,1]$, $a = 0$, $b = 1$, and define

$$\zeta(Y) = \begin{cases} (1-\lambda)(G-H)Y & \text{for } 0 \leqslant Y \leqslant \lambda \\ \lambda(G-H)(1-Y) & \text{for } \lambda \leqslant Y \leqslant 1 . \end{cases}$$

Setting $F = \lambda G + (1-\lambda)H$ we obtain from (2.2)

$$\lambda W(X,G) + (1-\lambda)W(X,H) \geqslant W(X,\lambda F+(1-\lambda)G), \tag{2.3}$$

so that $W(X,\cdot)$ is convex on $(0,\infty)$. ($\zeta \notin \mathfrak{D}(0,1)$ but we can nevertheless deduce (2.3) by approximation). Conversely, if $W(X,\cdot)$ is convex on $(0,\infty)$ and if $\zeta \in \mathfrak{D}(a,b)$ satisfies $F + \zeta'(Y) > 0$ for all $Y \in (a,b)$ then

$$W(F+\zeta'(Y)) \geqslant W(F) + A(F)\zeta'(Y)$$

for some $A(F) \in \mathbb{R}$, and (2.2) follows by integration.

In contrast to the situation in one dimension, for homogeneous materials quasiconvexity of W is not necessary for the existence of $C^1(\bar{\Omega})$ minimizers for 'all' displacement BVP's. For simplicity we give a two-dimensional example. Let $\Omega = \{\underset{\sim}{X} \in \mathbb{R}^2 : 1 < |\underset{\sim}{X}| < 2\}$. Let $W : M^{2\times 2} \longrightarrow \mathbb{R}$ be defined by $W(F) = \rho(r)$, where $r = |F| = [\operatorname{tr}(FF^T)]^{\frac{1}{2}}$, and where $\rho(r) = 0$ for $r \geq 1$, $\rho(r) > 0$ for $0 \leq r < 1$. Consider the map $\underset{\sim}{x}^{(n)}(\underset{\sim}{X})$ given in polar coordinates by $(R, \Theta) \longmapsto (R, \Theta + 2n\pi(R-1))$, where $R = |\underset{\sim}{X}|$ and $n = 1,2,\ldots$. Clearly $\underset{\sim}{x}^{(n)}(\underset{\sim}{X}) = \underset{\sim}{X}$ for $\underset{\sim}{X} \in \partial\Omega$ and $\det \nabla \underset{\sim}{x}^{(n)}(\underset{\sim}{X}) = 1$ for all $\underset{\sim}{X} \in \Omega$. Also it is easily checked that there are numbers a_n, $a_n \longrightarrow \infty$ as $n \longrightarrow \infty$, such that $|\nabla \underset{\sim}{x}^{(n)}(\underset{\sim}{X})| \geq a_n$ for all $\underset{\sim}{X} \in \bar{\Omega}$. Now let $\underset{\sim}{x}_o \in C^1(\bar{\Omega})$ satisfy $\det \nabla \underset{\sim}{x}_o(\underset{\sim}{X}) > 0$ for all $\underset{\sim}{X} \in \bar{\Omega}$. The map $\underset{\sim}{y}^{(n)} = \underset{\sim}{x}_o \circ \underset{\sim}{x}^{(n)}$ satisfies $\underset{\sim}{y}^{(n)}(\underset{\sim}{X}) = \underset{\sim}{x}_o(\underset{\sim}{X})$ for $\underset{\sim}{X} \in \partial\Omega$, $\det \nabla \underset{\sim}{y}^{(n)}(\underset{\sim}{X}) > 0$ for all $\underset{\sim}{X} \in \bar{\Omega}$. But

$$a_n \leq |\nabla \underset{\sim}{x}^{(n)}(\underset{\sim}{X})| = |\nabla \underset{\sim}{x}_o(\underset{\sim}{x}^{(n)}(\underset{\sim}{X}))^{-1} \nabla \underset{\sim}{y}^{(n)}(\underset{\sim}{X})| \leq C|\nabla \underset{\sim}{y}^{(n)}(\underset{\sim}{X})|$$

for all $\underset{\sim}{X} \in \Omega$ and some constant $C > 0$. Hence

$$\int_\Omega W(\nabla \underset{\sim}{y}^{(n)}(\underset{\sim}{X}))d\underset{\sim}{X} = 0$$

for large enough n, so that any displacement BVP has a $C^1(\bar{\Omega})$ minimizer. However W is not quasiconvex, as can be seen by setting $\underset{\sim}{x}_o(\underset{\sim}{X}) = \frac{1}{2}\underset{\sim}{X}$ in the above argument.

Curiously, if Ω is a cube quasiconvexity of W <u>is</u> a necessary condition for the existence of $C^1(\bar{\Omega})$ minimizers for certain displacement BVP's (see [A Thm 3.2]).

<u>Definition</u> W is <u>rank 1 convex</u> if for each $\underset{\sim}{X} \in \Omega$, $W(\underset{\sim}{X}, \cdot)$ is convex on all closed line segments in U with end points differing by a matrix of rank 1 i.e., if $\underset{\sim}{X} \in \Omega$ then

$$W(\underset{\sim}{X}, F + (1-\lambda)\, \underset{\sim}{a} \otimes \underset{\sim}{b}) \leq \lambda W(\underset{\sim}{X}, F) + (1-\lambda) W(\underset{\sim}{X}, F + \underset{\sim}{a} \otimes \underset{\sim}{b})$$

for all $F \in U$, $\lambda \in [0,1]$, $\underset{\sim}{a}, \underset{\sim}{b} \in \mathbb{R}^3$, with $F + \mu\, \underset{\sim}{a} \otimes \underset{\sim}{b} \in U$ for all $\mu \in [0,1]$. Here $(\underset{\sim}{a} \otimes \underset{\sim}{b})^i_\alpha \overset{\text{def}}{=} a^i b_\alpha$.

The geometric significance of matrices of rank 1 is the following: if $\underset{\sim}{x}(\underset{\sim}{X})$ is continuous and if $\nabla\underset{\sim}{x}$ takes the constant values F,G on opposite sides of the plane $\underset{\sim}{X}.\underset{\sim}{n} = k$, then $F - G = \underset{\sim}{\lambda}\otimes\underset{\sim}{n}$ for some $\underset{\sim}{\lambda} \in \mathbb{R}^3$.

The next theorem is a consequence of standard results on convex functions. For details see [A].

<u>Theorem 2.2</u> The following conditions (i), (ii) are equivalent

(i) W is rank 1 convex

(ii) for each $\underset{\sim}{X} \in \Omega, F \in U$ there exists $A(\underset{\sim}{X},F) \in M^{3\times 3}$ such that

$$W(\underset{\sim}{X},F + \underset{\sim}{a}\otimes\underset{\sim}{b}) \geqslant W(\underset{\sim}{X},F) + A_i^\alpha(\underset{\sim}{X},F)a^i b_\alpha$$

whenever $F + \lambda\, \underset{\sim}{a}\otimes\underset{\sim}{b} \in U$ for all $\lambda \in [0,1]$.

If $W(\underset{\sim}{X},\cdot)$ is C^2 for each $\underset{\sim}{X} \in \Omega$ then (i) and (ii) are equivalent to the <u>ellipticity</u> condition

(iii) $$\frac{\partial^2 W(\underset{\sim}{X},F)}{\partial F^i_\alpha \partial F^j_\beta} a^i a^j b_\alpha b_\beta \geqslant 0 \quad \text{for all } \underset{\sim}{X} \in \Omega,\ F \in U,\ \underset{\sim}{a},\underset{\sim}{b} \in \mathbb{R}^3.$$

If $W(\underset{\sim}{X},\cdot)$ is C^2 then the next result is simply Hadamard's theorem. For proofs of Hadamard's theorem see Graves [16] and Morrey [22,23]. A proof of Theorem 2.3 assuming only continuity of W is given in [A].

<u>Theorem 2.3</u> W quasiconvex implies W rank 1 convex.

Let $E = M^{3\times 3} \times M^{3\times 3} \times (0,\infty)$. We regard E as a subset of $\mathbb{R}^{19}$.

<u>Definition</u> W is <u>polyconvex</u> if there exists a function $g : \Omega \times E \longrightarrow \mathbb{R}$ such that

(i) $W(\underset{\sim}{X},F) = g(\underset{\sim}{X},F,\operatorname{adj} F, \det F)$ for all $\underset{\sim}{X} \in \Omega, F \in U$

(ii) $g(\underset{\sim}{X},\cdot,\cdot,\cdot)$ is convex on E.

Theorem 2.4 The following conditions (i) - (iii) are equivalent

(i) W is polyconvex

(ii) for each $\underset{\sim}{X} \in \Omega, F \in U$ there exist numbers $a_i^\alpha(\underset{\sim}{X},F)$, $b_i^\alpha(\underset{\sim}{X},F)$, $c(\underset{\sim}{X},F)$ such that

$$W(\underset{\sim}{X},\bar{F}) \geqslant W(\underset{\sim}{X},F) + a_i^\alpha(\bar{F}_\alpha^i - F_\alpha^i) + b_i^\alpha((\text{adj}\,\bar{F})_\alpha^i - (\text{adj}\,F)_\alpha^i) + c(\det\bar{F} - \det F)$$

for all $\bar{F} \in U$.

(iii) for each $\underset{\sim}{X} \in \Omega, F \in U$ there exist numbers $A_i^\alpha(\underset{\sim}{X},F)$, $B_i^\alpha(\underset{\sim}{X},F)$, $c(\underset{\sim}{X},F)$ such that

$$W(\underset{\sim}{X},F + \pi) \geqslant W(\underset{\sim}{X},F) + A_i^\alpha\pi_\alpha^i + B_i^\alpha(\text{adj}\,\pi)_\alpha^i + c(F)\det\pi$$

for all $F + \pi \in U$.

Proof. That (ii) and (iii) are equivalent follows immediately by setting $\bar{F} = F + \pi$ and rewriting the right hand sides of the inequalities. That (i) implies (ii) is a direct consequence of the convexity of $g(\underset{\sim}{X},\cdot,\cdot,\cdot)$. It remains to show that (ii) implies (i). Define g on $\Omega \times E$ by

$$g(\underset{\sim}{X},G,H,\delta) = \sup_{F\in U}\,[\,W(\underset{\sim}{X},F) + a_i^\alpha(\underset{\sim}{X},F)(G_\alpha^i - F_\alpha^i) + b_i^\alpha(\underset{\sim}{X},F)(H_\alpha^i - (\text{adj}\,F)_\alpha^i) + c(\underset{\sim}{X},F)(\delta - \det F)].$$

Fix $\underset{\sim}{X} \in \Omega$. As $g(\underset{\sim}{X},\cdot,\cdot,\cdot)$ is the supremum of a family of affine functions it is convex. By (ii),

$$g(\underset{\sim}{X},F,\text{adj}\,F,\det F) = W(\underset{\sim}{X},F), \qquad F \in U.$$

The only thing to check is that $g(\underset{\sim}{X},G,H,\delta) < \infty$ on $\Omega \times E$. Since $g(\underset{\sim}{X},\cdot,\cdot,\cdot)$ is convex it suffices to prove that the convex hull of the set $\{(F,\text{adj}\,F,\det F): F \in U\}$ is the whole of E. For $k > 0$, define $V_k \subset M^{3\times3} \times M^{3\times3}$ by

$$V_k = \{(F,\text{adj}\,F) : F \in M^{3\times3}, \det F = k\}.$$

It is enough to show that the convex hull of V_k is $M^{3\times3} \times M^{3\times3}$. Suppose not. Then (cf Rockafellar [32 p 99]) there is a closed half-space

$$\pi = \{(F,A) \in M^{3\times3} \times M^{3\times3} : F^i_\alpha G^\alpha_i + A^i_\alpha H^\alpha_i \leqslant \mu\},$$

$(G,H) \neq 0$, with $V_k \subset \pi$. If $R_1, R_2 \in M^{3\times3}$ are proper orthogonal then

$$F^i_\alpha G^\alpha_i + A^i_\alpha H^\alpha_i = \text{tr}\,[(R_1 F R_2)(R_2{}^T G R_1{}^T) + (R_2{}^T A R_1{}^T)(R_1 H R_2)].$$

Since $\text{adj}(R_1FR_2) = R_2{}^T(\text{adj}\,F)\,R_1{}^T$, $\det(R_1FR_2) = \det F$, we may without loss of generality suppose that H is diagonal. Suppose that $H \neq 0$ and assume without loss of generality that $H^1_1 \neq 0$. Let $F = \text{diag}(kN^{-1}\text{sgn}\,H^1_1, N^{\frac{1}{2}}\text{sgn}\,H^1_1, N^{\frac{1}{2}})$. Then $\text{adj}\,F = \text{diag}(N\ \text{sgn}\,H^1_1, kN^{-\frac{1}{2}}\text{sgn}\,H^1_1, kN^{-\frac{1}{2}})$ and $\det F = k$. Hence $(F, \text{adj}\,F) \in V_k$, but for $N > 0$ large enough $(F,\text{adj}\,F) \notin \pi$. If $H = 0$ then we can assume that $G^1_1 \neq 0$, let $F = \text{diag}(kN\,\text{sgn}\,G^1_1, N^{-\frac{1}{2}}\text{sgn}\,G^1_1, N^{-\frac{1}{2}})$ and proceed similarly. Thus $V_k \not\subset \pi$ and this contradiction completes the proof. □

Another equivalent condition for polyconvexity is given in [A] using work of Busemann, Ewald and Shephard [6] on convex functions defined on nonconvex sets.

Open problem

1. Give a physical interpretation of polyconvexity.

The following formulae, which express $\text{adj}\,\nabla\underset{\sim}{x}$ and $\det\nabla\underset{\sim}{x}$ as divergences, are fundamental to the rest of our work:

$$(\text{adj}\,\nabla\underset{\sim}{x})^\alpha_i \overset{\text{def}}{=} \tfrac{1}{2}\varepsilon_{ijk}\varepsilon^{\alpha\beta\gamma}\, x^j_{,\beta} x^k_{,\gamma} = (\tfrac{1}{2}\varepsilon_{ijk}\varepsilon^{\alpha\beta\gamma}\, x^j x^k_{,\gamma})_{,\beta} \tag{2.4}$$

$$\det\nabla\underset{\sim}{x} \overset{\text{def}}{=} \tfrac{1}{6}\varepsilon_{ijk}\varepsilon^{\alpha\beta\gamma}\, x^i_{,\alpha} x^j_{,\beta} x^k_{,\gamma} = (\tfrac{1}{3}\, x^i (\text{adj}\,\nabla\underset{\sim}{x})^\alpha_i)_{,\alpha}. \tag{2.5}$$

Clearly (2.4), (2.5) are valid if $\underset{\sim}{x}$ is C^2.

Theorem 2.5 (Morrey [22]) W polyconvex implies W quasiconvex.

Proof. Fix $\underset{\sim}{X} \in \Omega$. Let D be a bounded open subset of $\mathbb{R}^3$, let $F \in U$, and let $\underset{\sim}{\zeta} \in \mathcal{D}(D)$ satisfy $F + \nabla\underset{\sim}{\zeta}(\underset{\sim}{Y}) \in U$ for all $\underset{\sim}{Y} \in \Omega$. By (2.4) and (2.5),

$$\int_D \zeta^i{}_{,\alpha} d\underset{\sim}{Y} = \int_D (\operatorname{adj} \nabla\underset{\sim}{\zeta})^i_\alpha \, d\underset{\sim}{Y} = \int_D \det \nabla\underset{\sim}{\zeta} d\underset{\sim}{Y} = 0.$$

Thus by Theorem 2.4(iii),

$$\int_D W(\underset{\sim}{X}, F + \nabla\zeta(\underset{\sim}{Y}))d\underset{\sim}{Y} \geqslant \int_D W(\underset{\sim}{X}, F)d\underset{\sim}{Y}$$

as required. □

Let us call W convex if $W(\underset{\sim}{X}, \cdot)$ is convex on all closed line segments in U. Then we have the following situation:

$$W \text{ convex } \underset{\not\Leftarrow}{\Rightarrow} W \text{ polyconvex } \underset{\not\Leftarrow}{\Rightarrow} W \text{ quasiconvex } \underset{?}{\overset{\Rightarrow}{\Leftarrow}} W \text{ rank 1 convex}$$

$W(F) = \det F$ is an example for the first nonimplication; for the second see [A].

Open problems (see [A] for discussion)

2. Does W rank 1 convex imply W quasiconvex?
3. Does W quasiconvex imply W polyconvex if W satisfies (1.12)?

My guess is that the answer to both questions is no.

Consider next a pure displacement BVP with boundary condition $\underset{\sim}{x} = \bar{\underset{\sim}{x}}(\underset{\sim}{X})$ for $\underset{\sim}{X} \in \partial\Omega$, where $\bar{\underset{\sim}{x}} : \Omega \longrightarrow \mathbb{R}^3$ is globally one to one. Under assumptions like (1.15), one can hope that any minimizer $\underset{\sim}{x}$ for this problem will be globally one to one. Let $\underset{\sim}{X} = \underset{\sim}{x}^{-1}(\cdot)$, so that $\underset{\sim}{X} : \bar{\underset{\sim}{x}}(\Omega) \longrightarrow \Omega$. It seems reasonable to expect that $\underset{\sim}{X}$ will minimize

$$\hat{I}(\underset{\sim}{X}) = \int_{\bar{\underset{\sim}{x}}(\Omega)} \hat{W}(\underset{\sim}{X}(\underset{\sim}{x}),\nabla\underset{\sim}{X}(\underset{\sim}{x}))d\underset{\sim}{x} + \int_{\bar{\underset{\sim}{x}}(\Omega)} \psi(\underset{\sim}{x})\det\nabla\underset{\sim}{X}(\underset{\sim}{x})d\underset{\sim}{x}$$

in a suitable function space, subject to the boundary condition $\underset{\sim}{X}(\underset{\sim}{x}) = \bar{\underset{\sim}{x}}^{-1}(\underset{\sim}{x})$ for $\underset{\sim}{x} \in \partial\bar{\underset{\sim}{x}}(\Omega)$, where $\hat{W} : \Omega \times U \longrightarrow \mathbb{R}$ is defined by

$$\hat{W}(\underset{\sim}{X},F) = W(\underset{\sim}{X},F^{-1})\det F.$$

Note that ^ is an involution, i.e., $\hat{\hat{W}} = W$. We now ask which constitutive inequalities are invariant under the transformation $W \longmapsto \hat{W}$: it would be disconcerting if a constitutive inequality used as a hypothesis for an existence theorem did not possess this invariance, since then it might be possible to find a minimizer for I but not for $\hat{I}$ (or vice versa). The example $W \equiv 1$ shows that convexity of W is not invariant under ^.

Theorem 2.6 Quasiconvexity, rank 1 convexity, and polyconvexity are all invariant under ^.

Proof. Without loss of generality we take W independent of $\underset{\sim}{X}$. Let W be quasiconvex, D be a bounded open subset of $\mathbb{R}^3$, $\underset{\sim}{\zeta} \in \mathcal{D}(D)$, $F \in U$, and let $F + \nabla\underset{\sim}{\zeta}(\underset{\sim}{Y}) \in U$ for each $\underset{\sim}{Y} \in D$. Define $\underset{\sim}{x}(\underset{\sim}{Y}) = F\underset{\sim}{Y} + \underset{\sim}{\zeta}(\underset{\sim}{Y})$. Clearly $\det\nabla\underset{\sim}{x}(\underset{\sim}{Y}) \geqslant c > 0$ for all $\underset{\sim}{Y} \in D$. Also $\underset{\sim}{x}(\underset{\sim}{Y}) = F\underset{\sim}{Y}$ for $\underset{\sim}{Y} \in \partial D$, so that $\underset{\sim}{x}$ coincides on ∂D with a globally one to one function. Hence there exists an inverse function $\underset{\sim}{Y} : \underset{\sim}{x}(D) \longrightarrow D$, and $\underset{\sim}{Y}(\underset{\sim}{x}) = F^{-1}\underset{\sim}{x} + \underset{\sim}{\eta}(\underset{\sim}{x}) \in \mathcal{D}(\underset{\sim}{x}(D))$.

(While intuitively obvious, this step is not trivial and requires use of the Brouwer degree. The set $\underset{\sim}{x}(D)$ is open by the invariance of domain theorem.) Thus

$$\int_D \hat{W}(F + \nabla\underset{\sim}{\zeta}(\underset{\sim}{Y}))d\underset{\sim}{Y} = \int_{\underset{\sim}{x}(D)} W(F^{-1} + \nabla\eta(\underset{\sim}{x}))d\underset{\sim}{x} \geqslant \int_{\underset{\sim}{x}(D)} W(F^{-1})d\underset{\sim}{x} =$$

$$= W(F^{-1})\int_D \det(F + \nabla\underset{\sim}{\zeta}(\underset{\sim}{Y}))d\underset{\sim}{Y} = \int_D \hat{W}(F)d\underset{\sim}{Y},$$

where at the last step one has to use (2.4), (2.5). This proves that $\hat{W}$ is quasiconvex.

Let W be rank 1 convex. Let $F + \mu\, \underset{\sim}{a}\otimes\underset{\sim}{b} \in U$ for all $\mu \in [0,1]$. Since $(F + \underset{\sim}{a}\otimes\underset{\sim}{b})^{-1} - F^{-1} = -(F + \underset{\sim}{a}\otimes\underset{\sim}{b})^{-1}\, \underset{\sim}{a}\otimes\underset{\sim}{b}\, F^{-1}$ is of rank 1, it follows from Theorem 2.2(iii) that

$$\begin{aligned}\hat{W}(F + \underset{\sim}{a}\otimes\underset{\sim}{b}) &= W(F^{-1} + (F + \underset{\sim}{a}\otimes\underset{\sim}{b})^{-1} - F^{-1})\det(F + \underset{\sim}{a}\otimes\underset{\sim}{b}) \\ &\geqslant [\, W(F^{-1}) - A_i^{\alpha}(F)(F + \underset{\sim}{a}\otimes\underset{\sim}{b})^{-1}{}^{i}_{\beta}\, a^{\beta} b_j (F^{-1})^{j}_{\alpha}]\det(F + \underset{\sim}{a}\otimes\underset{\sim}{b}) \\ &= \hat{W}(F) + [(\operatorname{adj} F)(1 - F^{-1}A(F))]^{\alpha}_{i}(\underset{\sim}{a}\otimes\underset{\sim}{b})^{i}_{\alpha}.\end{aligned}$$

Thus $\hat{W}$ is rank 1 convex by Theorem 2.2.

Let W be polyconvex. Let $F, \bar{F} \in U$. By Theorem 2.4(ii),

$$\begin{aligned}\hat{W}(\bar{F}) \geqslant (&W(F^{-1}) + a_i^{\alpha}(F^{-1})(\bar{F}^{-1} - F^{-1})^{i}_{\alpha} + b_i^{\alpha}(F^{-1})[\operatorname{adj} \bar{F}^{-1})^{i}_{\alpha} - (\operatorname{adj} F^{-1})^{i}_{\alpha}] + \\ &+ c(F^{-1})[\det \bar{F}^{-1} - \det F^{-1}]\,)\det \bar{F} \\ = \hat{W}(F) &+ b_i^{\alpha}(F^{-1})(\bar{F} - F)^{i}_{\alpha} + a_i^{\alpha}(F^{-1})(\operatorname{adj} \bar{F} - \operatorname{adj} F)^{i}_{\alpha} - (\det F)^{-1}[\, W(F^{-1})\det F + \\ &+ a_i^{\alpha}(F^{-1})(\operatorname{adj} F)^{i}_{\alpha} + b_i^{\alpha}(F^{-1})F^{i}_{\alpha} + c(F^{-1})]\,(\det \bar{F} - \det F).\end{aligned}$$

Hence $\hat{W}$ is polyconvex by Theorem 2.4. □

The reader may find it interesting to see what Theorem 2.6 says in one dimension. Most of the results in this section can be carried over to the case of functions $\underset{\sim}{x} : \Omega \longrightarrow \mathbb{R}^n$, $\Omega \subset \mathbb{R}^m$, with little change in the proofs.

3 NULL LAGRANGIANS AND SEQUENTIALLY WEAKLY CONTINUOUS FUNCTIONS

Let Ω be a nonempty bounded open subset of $\mathbb{R}^3$.

<u>Definition</u> Let $\phi : M^{3\times 3} \longrightarrow \mathbb{R}$ be continuous. Then ϕ is said to be a <u>null Lagrangian</u> if

$$\int_\Omega \phi(\nabla \underset{\sim}{x}(\underset{\sim}{X}) + \nabla \underset{\sim}{\zeta}(\underset{\sim}{X}))d\underset{\sim}{X} = \int_\Omega \phi(\nabla \underset{\sim}{x}(\underset{\sim}{X}))d\underset{\sim}{X} \tag{3.1}$$

for all $\underset{\sim}{x} \in C^1(\bar{\Omega})$, $\underset{\sim}{\zeta} \in \mathfrak{D}(\Omega)$.

<u>Theorem 3.1</u> The following conditions (i) - (v) are equivalent

(i) ϕ is a null Lagrangian

(ii) $\int_\Omega \phi(F + \nabla\underset{\sim}{\zeta}(\underset{\sim}{X}))d\underset{\sim}{X} = \phi(F) \times \text{volume of } \Omega$

for all $F \in M^{3\times 3}$, $\underset{\sim}{\zeta} \in \mathfrak{D}(\Omega)$. (i.e., ϕ and $-\phi$ are quasiconvex on $M^{3\times 3}$)

(iii) ϕ is rank 1 affine, i.e.,

$$\phi(F + (1-\lambda)\underset{\sim}{a} \otimes \underset{\sim}{b}) = \lambda\phi(F) + (1-\lambda)\phi(F + \underset{\sim}{a} \otimes \underset{\sim}{b}) \tag{3.2}$$

for all $F \in M^{3\times 3}$, $\lambda \in [0,1]$, $\underset{\sim}{a},\underset{\sim}{b} \in \mathbb{R}^3$

(iv)
$$\phi(F) = A + A^\alpha_i F^i_\alpha + B^\alpha_i (\text{adj } F)^i_\alpha + C \det F \tag{3.3}$$

for all $F \in M^{3\times 3}$, where A, A^α_i, B^α_i, C are constants.

(v) ϕ is C^1 and

$$\int_\Omega \frac{\partial\phi}{\partial F^i_\alpha}(\nabla\underset{\sim}{x}(\underset{\sim}{X}))\zeta^i_{,\alpha}(\underset{\sim}{X})d\underset{\sim}{X} = 0 \tag{3.4}$$

for all $\underset{\sim}{x} \in C^1(\bar{\Omega})$, $\underset{\sim}{\zeta} \in \mathfrak{D}(\Omega)$.

Proof. Putting $\underset{\sim}{x}(\underset{\sim}{X}) = F\underset{\sim}{X}$ in (3.1) shows that (i) implies (ii). Let (ii) hold. Then, by an obvious modification of Theorem 2.3, ϕ and $-\phi$ are rank 1 convex on $M^{3\times 3}$, i.e., (iii) holds. To show that (iii) implies (iv) assume first that ϕ is C^2, so that (c.p. Theorem 2.2 (iii))

$$\frac{\partial^2\phi(F)}{\partial F^i_\alpha \partial F^j_\beta} a^i a^j b_\alpha b_\beta = 0 \quad \text{for all } F \in M^{3\times 3}, \quad \underset{\sim}{a},\underset{\sim}{b} \in \mathbb{R}^3. \tag{3.5}$$

A result of Ericksen [11] shows that (3.5) holds if and only if ϕ has the form (3.3). (See also Edelen [9]). For a general continuous ϕ one can use a mollifier argument to reduce the problem to the case $\phi \in C^2$; for details see [A].

To prove that (iv) implies (v), note first that by approximation we may without loss of generality assume that $\underset{\sim}{x} \in C^2(\Omega)$. Then, if $\underset{\sim}{\zeta} \in \mathcal{D}(\Omega)$,

$$\int_\Omega \frac{\partial\phi}{\partial F^i_\alpha}(\nabla \underset{\sim}{x}(\underset{\sim}{X}))\zeta^i_{,\alpha}(\underset{\sim}{X})d\underset{\sim}{X} = -\int_\Omega \frac{\partial^2\phi}{\partial F^i_\alpha \partial F^j_\beta}(\nabla \underset{\sim}{x}(\underset{\sim}{X}))x^j_{,\beta\alpha}\,\zeta^i d\underset{\sim}{X}.$$

But (3.5) implies that $\frac{\partial^2\phi}{\partial F^i_\alpha \partial F^j_\beta} = -\frac{\partial^2\phi}{\partial F^i_\beta \partial F^j_\alpha}$. This gives (v).

Finally let (v) hold and let $\underset{\sim}{x} \in C^1(\Omega)$, $\underset{\sim}{\zeta} \in \mathcal{D}(\Omega)$. For $t \in [0,1]$ define

$$g(t) = \int_\Omega \phi(\nabla \underset{\sim}{x}(\underset{\sim}{X}) + t\nabla\underset{\sim}{\zeta}(\underset{\sim}{X}))d\underset{\sim}{X}.$$

Then, by (v), $g'(t) \equiv 0$. Hence $g(1) = g(0)$, which is (i). □

Theorem 3.2 Let $\phi : M^{3\times 3} \longrightarrow \mathbb{R}$ be continuous. Let $p \geqslant 1$ and suppose that $|\phi(F)| \leqslant C(1 + |F|^p)$ for some constant $C \geqslant 0$ and all $F \in M^{3\times 3}$. If the map $\underset{\sim}{x} \longmapsto \phi(\nabla\underset{\sim}{x}(\cdot)) : W^{1,p}(\Omega) \longrightarrow L^1(\Omega)$ is sequentially weakly continuous then ϕ is a null Lagrangian.

Proof We use an argument of Morrey [22]. Let D be the unit cube $0 < X^\alpha < 1$. Let $F \in M^{3\times 3}$, $\underset{\sim}{\zeta} \in \mathcal{D}(D)$, $\varepsilon > 0$. Tesselate $\mathbb{R}^3$ by cubes of side ε with faces perpendicular to the X^α axes. Let $D^{(r)} = \underset{\sim}{X}^{(r)} + \varepsilon D$ be a typical cube, and define $\underset{\sim}{x}_\varepsilon : \Omega \longrightarrow \mathbb{R}^3$ by

$$\underset{\sim}{x}_\varepsilon(\underset{\sim}{X}) = \begin{cases} F\underset{\sim}{X} + \varepsilon\,\underset{\sim}{\zeta}\left(\dfrac{\underset{\sim}{X}-\underset{\sim}{X}^{(r)}}{\varepsilon}\right) & \text{if } \underset{\sim}{X} \in D^{(r)} \subset \Omega \\ F\underset{\sim}{X} & \text{otherwise.} \end{cases}$$

Then

$$\nabla \underset{\sim}{x}_\varepsilon(\underset{\sim}{X}) = \begin{cases} F + \nabla\underset{\sim}{\zeta}\left(\dfrac{\underset{\sim}{X}-\underset{\sim}{X}^{(r)}}{\varepsilon}\right) & \text{if } \underset{\sim}{X} \in D^{(r)} \subset \Omega \\ F & \text{otherwise.} \end{cases}$$

Hence $|\nabla \underset{\sim}{x}_\varepsilon(\underset{\sim}{X})|$ is uniformly bounded for all $\underset{\sim}{X}$ and ε. Since $\underset{\sim}{x}_\varepsilon(\underset{\sim}{X}) \longrightarrow F\underset{\sim}{X}$ uniformly in $\bar{\Omega}$ as $\varepsilon \longrightarrow 0$, it follows that $\underset{\sim}{x}_\varepsilon(\cdot) \longrightarrow F\underset{\sim}{X}$ in $W^{1,p}(\Omega)$ as $\varepsilon \longrightarrow 0$. But if $D^{(r)} \subset \Omega$ then

$$\int_{D^{(r)}} \phi(\nabla \underset{\sim}{x}_\varepsilon(\underset{\sim}{X}))d\underset{\sim}{X} = \varepsilon^3 \int_D \phi(F + \nabla\underset{\sim}{\zeta}(\underset{\sim}{Y}))d\underset{\sim}{Y}.$$

As the number of cubes $D^{(r)} \subset \Omega$ is of order $\dfrac{1}{\varepsilon^3}$ X volume of Ω we obtain

$$\lim_{\varepsilon \to 0} \int_\Omega \phi(\nabla \underset{\sim}{x}_\varepsilon(\underset{\sim}{X}))d\underset{\sim}{X} = \int_D \phi(F + \nabla\underset{\sim}{\zeta}(\underset{\sim}{Y}))d\underset{\sim}{Y} \times \text{volume of } \Omega.$$

Since $\phi(\nabla \underset{\sim}{x}(\cdot))$ is sequentially weakly continuous it follows that

$$\int_D \phi(F + \nabla\underset{\sim}{\zeta}(\underset{\sim}{Y}))d\underset{\sim}{Y} = \int_D \phi(F)d\underset{\sim}{Y}.$$

Hence ϕ is a null Lagrangian by Theorem 3.1. □

The study of sufficient conditions for sequential weak continuity hinges on the identities (2.4), (2.5). First we give a distributional meaning to these identities. (For the definition of the dual space $\mathscr{D}'(\Omega)$ see Schwartz [33].)

<u>Lemma 3.3</u> (a) If $\underset{\sim}{x} \in W^{1,2}(\Omega)$ then adj $\nabla \underset{\sim}{x} \in L^1(\Omega)$ and formula (2.4) holds in $\mathscr{D}'(\Omega)$. (b) If $\underset{\sim}{x} \in W^{1,p}(\Omega)$, $p \geqslant 2$, and adj $\nabla \underset{\sim}{x} \in L^{p'}(\Omega)$, (where $\frac{1}{p}+\frac{1}{p'} = 1$) then det $\nabla \underset{\sim}{x} \in L^1(\Omega)$ and formula (2.5) holds in $\mathscr{D}'(\Omega)$.

<u>Proof.</u> (a) Let $\underset{\sim}{x} \in W^{1,2}(\Omega)$. Clearly adj $\nabla \underset{\sim}{x} \in L^1(\Omega)$. Formula (2.4) holds in $\mathscr{D}'(\Omega)$ if and only if

$$\int_\Omega (\text{adj}\, \nabla \underset{\sim}{x})^\alpha_i \phi \, d\underset{\sim}{X} = - \int_\Omega \tfrac{1}{2} \varepsilon_{ijk} \varepsilon^{\alpha\beta\gamma} x^j x^k_{,\gamma} \phi_{,\beta} \, d\underset{\sim}{X} \quad \text{for all } \phi \in \mathscr{D}(\Omega). \tag{3.6}$$

But (3.6) holds trivially if $\underset{\sim}{x} \in C^\infty(\Omega)$, and $C^\infty(\Omega)$ is norm dense in $W^{1,2}(\Omega)$. Since both sides of (3.6) are continuous functions of $\underset{\sim}{x} \in W^{1,2}(\Omega)$ the result follows.

(b) Let $\underset{\sim}{x} \in W^{1,p}(\Omega)$, adj $\nabla \underset{\sim}{x} \in L^{p'}(\Omega)$. Then det $\nabla \underset{\sim}{x} \in L^1(\Omega)$ by Hölder's inequality. For fixed i define $\underset{\sim}{w}_{(i)}$ by $w^\alpha_{(i)} = (\text{adj}\, \nabla \underset{\sim}{x})^\alpha_i$. If $\underset{\sim}{x} \in C^\infty(\Omega)$ then Div $\underset{\sim}{w}_{(i)} \overset{\text{def}}{=} w^\alpha_{(i),\alpha} = 0$. Thus

$$\int_\Omega w^\alpha_{(i)} \phi_{,\alpha} \, d\underset{\sim}{X} = 0 \qquad \text{for all } \phi \in \mathscr{D}(\Omega). \tag{3.7}$$

Since $C^\infty(\Omega)$ is dense in $W^{1,p}(\Omega)$ and $p \geqslant 2$ it follows that (3.7) holds for $\underset{\sim}{x} \in W^{1,p}(\Omega)$. Let

$$\rho \in \mathscr{D}(\mathbb{R}^3), \ \rho \geqslant 0, \int_{\mathbb{R}^3} \rho(X) dX = 1,$$

and define the mollifier ρ_k in the usual way by $\rho_k(\underset{\sim}{X}) = k^3 \rho(k\underset{\sim}{X})$. Extend $\underset{\sim}{w}_{(i)}$ by zero outside Ω. Then the convolution $\rho_k * \underset{\sim}{w}_{(i)} \in C^\infty(\mathbb{R}^3)$ and

$\rho_k * \underset{\sim}{w}_{(i)} \longrightarrow \underset{\sim}{w}_{(i)}$ in $L^{p'}(\mathbb{R}^3)$ as $k \longrightarrow \infty$. Fix $\rho \in \mathcal{D}(\Omega)$. If k is large enough, then by (3.7),

$$\operatorname{Div}(\rho_k * \underset{\sim}{w}_{(i)})(\underset{\sim}{X}) = \int_{\mathbb{R}^3} \rho_{k,\alpha}(\underset{\sim}{X}-\underset{\sim}{Y}) w^{\alpha}_{(i)}(\underset{\sim}{Y}) d\underset{\sim}{Y} = 0 \tag{3.8}$$

for all $\underset{\sim}{X} \in \operatorname{supp} \phi$. Let $S \subset \mathbb{R}^3$ be an open ball containing supp ϕ, and let $\underset{\sim}{x}_{(k)} \in C^{\infty}(\Omega)$, $\underset{\sim}{x}_{(k)} \longrightarrow \underset{\sim}{x}$ in $W^{1,p}(\Omega)$. Then, using (3.8),

$$\int_{\Omega} \tfrac{1}{3} x^{i}_{(k),\alpha} (\rho_k * \underset{\sim}{w}_{(i)})^{\alpha} \phi \, dX = \int_{S} \operatorname{Div} [\tfrac{1}{3} x^{i}_{(k)} (\rho_k * \underset{\sim}{w}_{(i)}) \phi] \, d\underset{\sim}{X}$$

$$- \int_{\Omega} \tfrac{1}{3} x^{i}_{(k)} (\rho_k * \underset{\sim}{w}_{(i)})^{\alpha} \phi_{,\alpha} \, d\underset{\sim}{X}$$

$$= - \int_{\Omega} \tfrac{1}{3} x^{i}_{(k)} (\rho_k * \underset{\sim}{w}_{(i)})^{\alpha} \phi_{,\alpha} \, d\underset{\sim}{X}.$$

Let $k \longrightarrow \infty$. Then

$$\int_{\Omega} \tfrac{1}{3} x^{i}_{,\alpha} w^{\alpha}_{(i)} \phi \, d\underset{\sim}{X} = - \int_{\Omega} \tfrac{1}{3} x^{i} w^{\alpha}_{(i)} \phi_{,\alpha} \, d\underset{\sim}{X},$$

which is (2.5). □

The main result of this section is

<u>Theorem 3.4</u> (a) Let $p \geqslant 2$. If $\underset{\sim}{x}_{(r)} \rightharpoonup \underset{\sim}{x}$ in $W^{1,p}(\Omega)$ then $\operatorname{adj} \nabla \underset{\sim}{x}_{(r)} \longrightarrow \operatorname{adj} \nabla \underset{\sim}{x}$ in $\mathcal{D}'(\Omega)$. (b) Let $p \geqslant 2$. If $\underset{\sim}{x}_{(r)} \rightharpoonup \underset{\sim}{x}$ in $W^{1,p}(\Omega)$ and $\operatorname{adj} \nabla \underset{\sim}{x}_{(r)} \rightharpoonup \operatorname{adj} \nabla \underset{\sim}{x}$ in $L^{p'}(\Omega)$, then $\det \nabla \underset{\sim}{x}_{(r)} \longrightarrow \det \nabla \underset{\sim}{x}$ in $\mathcal{D}'(\Omega)$.

<u>Proof</u>. (a) Fix $\phi \in \mathcal{D}(\Omega)$. By Lemma 3.3(a), for each r we have

$$\int_{\Omega} (\operatorname{adj} \nabla \underset{\sim}{x}_{(r)})^{\alpha}_{i} \phi \, d\underset{\sim}{X} = - \int_{\Omega} \tfrac{1}{2} \varepsilon_{ijk} \varepsilon^{\alpha\beta\gamma} x^{j}_{(r)} x^{k}_{(r),\gamma} \phi_{,\beta} \, d\underset{\sim}{X}.$$

Let Ω' be an open set with $\Omega \supset \Omega' \supset \text{supp}\,\phi$ and such that Ω' satisfies the cone condition (cf Adams [1]). Since $\underset{\sim}{x}_{(r)} \rightharpoonup \underset{\sim}{x}$ in $W^{1,p}(\Omega')$ it follows by the Rellich-Kondrachov theorem that $\underset{\sim}{x}_{(r)} \longrightarrow \underset{\sim}{x}$ in $L^2(\Omega')$. Hence $x^j_{(r)}\, x^k_{(r),\gamma} \rightharpoonup x^j x^k_{,\gamma}$ in $L^1(\Omega')$, so that

$$\int_\Omega (\text{adj}\,\nabla \underset{\sim}{x}_{(r)})^\alpha_i\, \phi\, d\underset{\sim}{X} \longrightarrow \int_\Omega (\text{adj}\,\nabla \underset{\sim}{x})^\alpha_i\, \phi\, d\underset{\sim}{X}.$$

(b) This is proved similarly using Lemma 3.3(b). □

Corollary 3.5 (Reshetnyak [27,28]) (a) If $p > 2$ the map $\underset{\sim}{x} \longmapsto L^{p/2}(\Omega)$ is sequentially weakly continuous. (b) If $p > 3$ the map $\underset{\sim}{x} \longmapsto \det \nabla \underset{\sim}{x} : W^{1,p}(\Omega) \longrightarrow L^{p/3}(\Omega)$ is sequentially weakly continuous.

Proof. (a) If $\underset{\sim}{x}_{(r)} \rightharpoonup \underset{\sim}{x}$ in $W^{1,p}(\Omega)$ then $\text{adj}\,\nabla \underset{\sim}{x}_{(r)}$ is bounded in $L^{p/2}(\Omega)$. Therefore a subsequence $\text{adj}\,\nabla \underset{\sim}{x}_{(\mu)} \rightharpoonup H$ in $L^{p/2}(\Omega)$. By part (a) of the theorem $H = \text{adj}\,\nabla \underset{\sim}{x}$. Hence the whole sequence converges weakly to $\text{adj}\,\nabla \underset{\sim}{x}$.
(b) If $\underset{\sim}{x}_{(r)} \rightharpoonup \underset{\sim}{x}$ in $W^{1,p}(\Omega)$ then we have just shown that $\text{adj}\,\nabla \underset{\sim}{x}_{(r)} \rightharpoonup \text{adj}\,\nabla \underset{\sim}{x}$ in $L^{p/2}(\Omega)$. Also $\det \nabla \underset{\sim}{x}_{(r)}$ is bounded in $L^{p/3}(\Omega)$, so that a subsequence $\det \nabla \underset{\sim}{x}_{(\mu)} \rightharpoonup \delta$ in $L^{p/3}(\Omega)$. By part (b) of the theorem $\delta = \det \nabla \underset{\sim}{x}$. Thus $\det \nabla \underset{\sim}{x}_{(r)} \rightharpoonup \det \nabla \underset{\sim}{x}$ in $L^{p/3}(\Omega)$. □

An amusing illustration of the sequential weak continuity of $\det \nabla \underset{\sim}{x}$ can be given in two dimensions. If Ω is a bounded open subset of $\mathbb{R}^2$, and we consider functions $\underset{\sim}{x} : \Omega \longrightarrow \mathbb{R}^2$, then identical arguments to the above show that $\underset{\sim}{x} \longmapsto \det \nabla \underset{\sim}{x} : W^{1,p}(\Omega) \longrightarrow L^{p/2}(\Omega)$ is sequentially weakly continuous for $p > 2$. Take Ω to be the unit square, and consider the sequence of maps $\underset{\sim}{x}_{(r)} : \Omega \longrightarrow \mathbb{R}^2$ shown in Figure 7 obtained by folding Ω into four along the dotted lines, into four again, and so on, keeping the origin fixed. Clearly $|\nabla \underset{\sim}{x}_{(r)}{}^i_\alpha(\underset{\sim}{X})| = 1$ for all r, i, α and almost all $\underset{\sim}{X} \in \Omega$. Hence some subsequence $\underset{\sim}{x}_{(\mu)} \rightharpoonup \underset{\sim}{x}$ in $W^{1,p}(\Omega)$

for any $p > 1$. Obviously $\underset{\sim}{x}_{(r)} \longrightarrow \underset{\sim}{0}$ in $L^\infty(\Omega)$. Hence $\underset{\sim}{x}_{(r)} \longrightarrow \underset{\sim}{0}$ in $W^{1,p}(\Omega)$. The sequence $\det \nabla \underset{\sim}{x}_{(r)}$ must therefore converge weakly to zero in any space $L^p(\Omega)$. That this is indeed the case can be seen from Figure 7, where $\det \nabla \underset{\sim}{x}_{(r)}$ takes the values +1, -1 in the unshaded and shaded regions respectively.

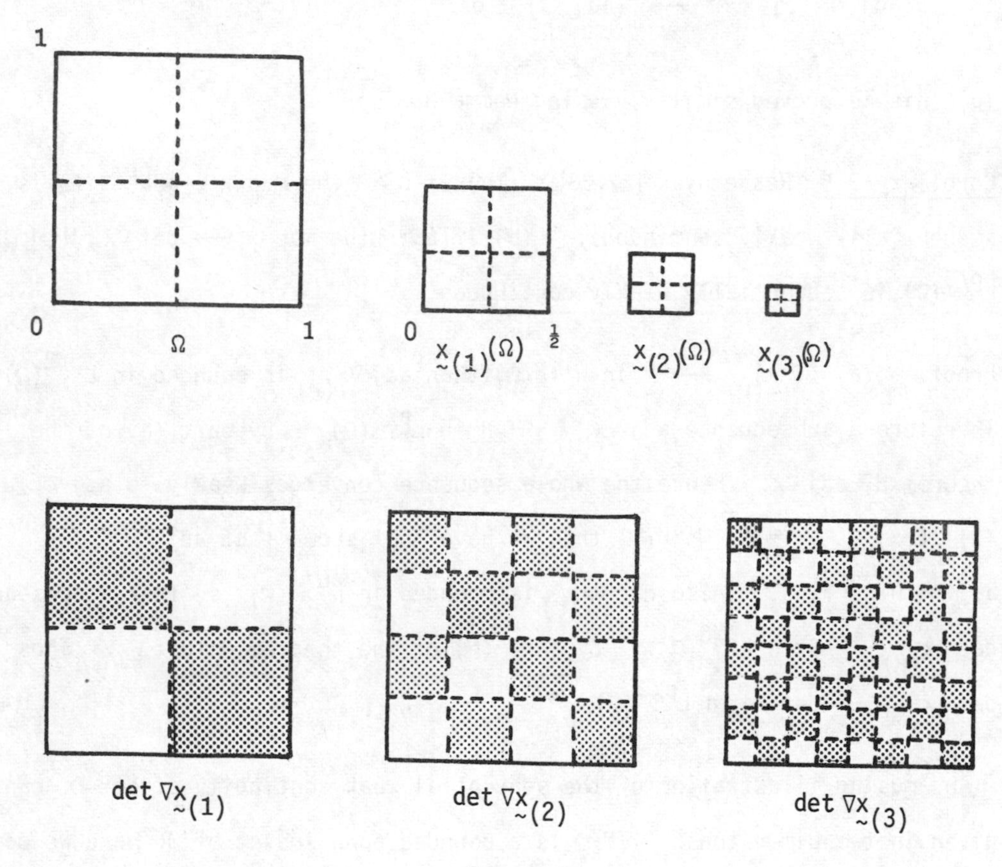

Figure 7

The results of this section carry over in a natural way to the case of maps $\underset{\sim}{x} : \mathbb{R}^m \longrightarrow \mathbb{R}^n$, $m,n \geqslant 1$. See Ball [5] for details.

4 EXISTENCE OF MINIMIZERS

We confine attention to mixed displacement zero traction BVP's. A variety of other boundary conditions are treated using similar methods in [A].

We make the following hypotheses on the stored-energy function $W:\Omega\times U \longrightarrow \mathbb{R}$.

(H1) W is polyconvex, and the corresponding function $g:\Omega\times E \longrightarrow \mathbb{R}$ is continuous.

(H2) There exist constants $K>0$, C, $p\geqslant 2$, $q\geqslant\frac{p}{p-1}$, $r>1$, such that

$$g(\underset{\sim}{X},F,H,\delta)\geqslant C+K(|F|^p+|H|^q+|\delta|^r)$$

for all $\underset{\sim}{X}\in\Omega$, $(F,H,\delta)\in E$.

(H3) $g(\underset{\sim}{X},a)\longrightarrow\infty$ as $a\longrightarrow\partial E$.

Remarks: The continuity hypothesis on g may be weakened (see [A]). (H3) corresponds to (1.15).

We define $g(\underset{\sim}{X},a)$ to be $+\infty$ if $a\in\partial E$, so that by (H3) $g:\Omega\times\bar{E}\longrightarrow\bar{\mathbb{R}}$ is continuous. We suppose that the body force potential $\psi:\mathbb{R}^3\longrightarrow\mathbb{R}^+$ and is continuous, that the boundary $\partial\Omega$ of Ω satisfies a strong Lipschitz condition, and that $\partial\Omega_1$ is a measurable subset of $\partial\Omega$ with positive measure. We seek a minimizer for

$$I(\underset{\sim}{x})=\int_\Omega W(\underset{\sim}{X},\nabla\underset{\sim}{x}(\underset{\sim}{X}))d\underset{\sim}{X}+\int_\Omega\psi(\underset{\sim}{x}(\underset{\sim}{X}))d\underset{\sim}{X} \tag{4.1}$$

subject to

$$\underset{\sim}{x}(\underset{\sim}{X})=\bar{\underset{\sim}{x}}(\underset{\sim}{X})\qquad\text{for } X\in\partial\Omega_1, \tag{4.2}$$

where $\bar{\underset{\sim}{x}}:\partial\Omega_1\longrightarrow\mathbb{R}^3$ is a given measurable function. We require (4.2) to be satisfied in the sense of trace.

Define the admissibility set A by

$$A = \{\underset{\sim}{x} \in W^{1,p}(\Omega); \text{adj } \nabla\underset{\sim}{x} \in L^q(\Omega), \det \nabla\underset{\sim}{x} \in L^r(\Omega), \det \nabla\underset{\sim}{x} > 0 \text{ almost everywhere in } \Omega, \underset{\sim}{x} = \bar{\underset{\sim}{x}} \text{ almost everywhere in } \partial\Omega_1\}.$$

Theorem 4.1 Suppose there exists $\underset{\sim}{x}_1 \in A$ with $I(\underset{\sim}{x}_1) < \infty$. Then there exists $\underset{\sim}{x}_o \in A$ which minimizes I in A.

Proof. Since $\partial\Omega_1$ has positive measure, a result in Morrey [23 p 82] implies that there exists $k_1 > 0$ such that

$$\int_\Omega |\underset{\sim}{x}|^p \, d\underset{\sim}{X} \leqslant k_1 \left[\int_\Omega |\nabla\underset{\sim}{x}|^p \, dX + \left(\int_{\partial\Omega_1} |\bar{\underset{\sim}{x}}| \, dS\right)^p\right] \tag{4.3}$$

for all $\underset{\sim}{x} \in W^{1,p}(\Omega)$ with $\underset{\sim}{x} = \bar{\underset{\sim}{x}}$ on $\partial\Omega_1$. Hence by (H2) we have for arbitrary $\underset{\sim}{x} \in A$,

$$I(\underset{\sim}{x}) \geqslant C \text{ (volume of } \Omega) + K\left[\int_\Omega |\nabla\underset{\sim}{x}|^p \, d\underset{\approx}{X} + \int_\Omega |\text{adj } \nabla\underset{\sim}{x}|^q \, d\underset{\sim}{X} + \int_\Omega (\det \nabla\underset{\sim}{x})^r \, d\underset{\sim}{X}\right]$$

$$\geqslant \text{const.} + K_1 \|\underset{\sim}{x}\|^p_{W^{1,p}(\Omega)} + K\left(\int_\Omega |\text{adj } \nabla\underset{\sim}{x}|^q \, d\underset{\sim}{X} + \int_\Omega (\det \nabla\underset{\sim}{x})^r \, d\underset{\sim}{X}\right), \tag{4.4}$$

where $K_1 > 0$ is a constant.

Let $\underset{\sim}{x}_{(n)}$ be a minimizing sequence for I in A. It follows from (4.4) that some subsequence $\underset{\sim}{x}_{(\mu)}$ satisfies

$$\underset{\sim}{x}_{(\mu)} \longrightarrow \underset{\sim}{x}_o \text{ in } W^{1,p}(\Omega), \; \underset{\sim}{x}_{(\mu)} \longrightarrow \underset{\sim}{x}_o \text{ almost everywhere in } \Omega \text{ and } \partial\Omega_1,$$

$$\text{adj } \nabla\underset{\sim}{x}_{(\mu)} \longrightarrow H \text{ in } L^q(\Omega), \; \det \nabla\underset{\sim}{x}_{(\mu)} \longrightarrow \delta \text{ in } L^r(\Omega).$$

By Theorem 3.4, $H = \operatorname{adj} \nabla \underset{\sim}{x}_o$ and $\delta = \det \nabla \underset{\sim}{x}_o$. Hence,

$$(\nabla \underset{\sim}{x}_{(\mu)},\ \operatorname{adj} \nabla \underset{\sim}{x}_{(\mu)},\ \det \nabla \underset{\sim}{x}_{(\mu)}) \longrightarrow (\nabla \underset{\sim}{x}_o,\ \operatorname{adj} \nabla \underset{\sim}{x}_o,\ \det \nabla \underset{\sim}{x}_o) \text{ in } L^1(\Omega).$$

Since $g(\underset{\sim}{X},\cdot,\cdot,\cdot)$ is convex, it follows by a lower semicontinuity theorem of Ekeland and Témam [10 Thm 2.1 p 226] that

$$I(\underset{\sim}{x}_o) \leqslant \varliminf_{\mu\to\infty} I(\underset{\sim}{x}_{(\mu)}).$$

But clearly $\underset{\sim}{x}_o = \bar{\underset{\sim}{x}}_o$ on $\partial\Omega_1$. Also, as $I(\underset{\sim}{x}_o) < \infty$ we must have $\det \nabla \underset{\sim}{x}_o > 0$ almost everywhere. Hence $\underset{\sim}{x}_o \in A$. This completes the proof. □

Incompressible elasticity

We retain the same hypotheses on Ω, $\partial\Omega_1$, ψ and $\bar{\underset{\sim}{x}}$, but replace (H1) - (H3) by hypotheses (H1)', (H2)' below.

Let $V = \{F \in M^{3\times3} : \det F = 1\}$.

(H1)' $W : \Omega \times V \longrightarrow \mathbb{R}$ and there exists a continuous function $g : \Omega \times M^{3\times3} \times M^{3\times3} \longrightarrow \mathbb{R}$, with $g(\underset{\sim}{X},\cdot,\cdot)$ convex, such that

$$W(\underset{\sim}{X},F) = g(\underset{\sim}{X},F, \operatorname{adj} F) \quad \text{for all } \underset{\sim}{X} \in \Omega, F \in V.$$

(H2)' There exist constants $K > 0$, C, $p \geqslant 2$, $q \geqslant p/p-1$, such that

$$g(\underset{\sim}{X},F,H) \geqslant C + K(|F|^p + |H|^q) \quad \text{for all } \underset{\sim}{X} \in \Omega,\ F,H \in M^{3\times3}.$$

Let $A' = \{\underset{\sim}{x} \in W^{1,p}(\Omega) : \operatorname{adj} \nabla \underset{\sim}{x} \in L^q(\Omega), \det \nabla \underset{\sim}{x} = 1$ almost everywhere in Ω, $\underset{\sim}{x} = \bar{\underset{\sim}{x}}$ almost everywhere in $\partial\Omega_1\}$.

Theorem 4.2 Suppose there exists $\underset{\sim}{x}_1 \in A'$ with $I(\underset{\sim}{x}_1) < \infty$. Then there exists $\underset{\sim}{x}_o \in A'$ which minimizes I on A'.

Proof. Let $\underset{\sim}{x}_{(n)}$ be a minimizing sequence for I from A'. Since $\det \nabla \underset{\sim}{x}_{(n)} = 1$ almost everywhere in Ω, we have in particular that $\det \nabla \underset{\sim}{x}_{(n)}$ is bounded in

$L^2(\Omega)$, say. Proceeding as in the proof of Theorem 4.1, we obtain a minimizing sequence $\underset{\sim}{x}_{(\mu)} \subset A'$ with the properties

$$\underset{\sim}{x}_{(\mu)} \longrightarrow \underset{\sim}{x}_o \text{ in } W^{1,p}(\Omega),\ \underset{\sim}{x}_{(\mu)} \longrightarrow \underset{\sim}{x}_o \text{ almost everywhere in } \Omega \text{ and } \partial\Omega_1,$$

$$\text{adj}\,\nabla\underset{\sim}{x}_{(\mu)} \longrightarrow \text{adj}\,\nabla\underset{\sim}{x}_o \text{ in } L^q(\Omega),\ \det\nabla\underset{\sim}{x}_{(\mu)} \longrightarrow \det\nabla\underset{\sim}{x}_o \text{ in } L^2(\Omega).$$

Thus $\det\nabla\underset{\sim}{x}_o = 1$ almost everywhere in Ω. Hence $\underset{\sim}{x}_o \in A'$ and we obtain the theorem as before. □

Note that in the proof of Theorem 4.2 we made essential use of the fact that the pointwise constraint $\det F = 1$ was weakly closed. The only other homogeneous constraints of this type, as we have seen in Theorem 3.2, are of the form

$$\phi(F) = A + A_i^\alpha F_\alpha^i + B_i^\alpha (\text{adj}\,F)_\alpha^i + D\det F = 0,$$

where A, A_i^α, B_i^α and D are constants. One can easily show that the only <u>frame-indifferent</u> constraints of this form (i.e., ϕ satisfying $\phi(QF) = \phi(F)$ for all proper orthogonal Q) are those with $A_i^\alpha = B_i^\alpha = 0$, so that $\det F$ is specified. Note that the constraint of inextensibility (Truesdell and Noll [38 p 72]) is not included. This makes one wonder about the mathematical status of inextensible elasticity.

Note also that if $\partial\Omega_1 = \partial\Omega$ and $\bar{\underset{\sim}{x}} \in C^1(\bar{\Omega})$ then a necessary condition for A' to be nonempty is that

$$\int_\Omega \det\nabla\bar{\underset{\sim}{x}}(\underset{\sim}{X})\,d\underset{\sim}{X} = \text{volume of } \Omega. \tag{4.5}$$

<u>Open problem</u>

4. Let $\bar{\underset{\sim}{x}}$ be a diffeomorphism satisfying (4.5). Does there exist a volume-preserving diffeomorphism $\underset{\sim}{x}$ with $\underset{\sim}{x} = \bar{\underset{\sim}{x}}$ on $\partial\Omega$?

5 REMARKS ON REGULARITY AND STORED-ENERGY FUNCTIONS OF SLOW GROWTH

Open problem

5. Prove that under suitable hypotheses the minimizers in Theorems 4.1 and 4.2 are smooth.

A necessary prerequisite for solving the regularity problem is presumably to show that the minimizers are weak solutions of the equilibrium equations. In the case of Theorem 4.1 the problem is very delicate due to the pointwise constraint $\det F > 0$ and the associated growth condition (H3). Even the simpler case of Theorem 4.2 presents serious difficulties; formally one could regard $\det \nabla \underset{\sim}{x} = 1$ as a Banach space valued constraint and apply an appropriate Lagrange multiplier theorem, identifying the Lagrange multiplier with the familiar hydrostatic pressure of incompressible elasticity. However, satisfying the hypotheses of standard Lagrange multiplier theorems is not straightforward because it is not a priori obvious that the minimizer $\underset{\sim}{x}_o$ is invertible, even in the case of a pure displacement BVP. One could minimize in a class of invertible functions, but then other difficulties arise. The only result on weak solutions I know of avoids these problems by assuming that they have already been overcome.

In addition to the hypotheses of Theorem 4.1 we suppose that $p \geq q \geq r$ and

(H4) $W(\underset{\sim}{X}, \cdot) \in C^1(U)$ for each $\underset{\sim}{X} \in \Omega$, and for each $d > 0$ there exists a constant $C(d)$ such that

$$\left| \frac{\partial W(\underset{\sim}{X}, F)}{\partial F} \right| \leq C(d)\,(1 + |F|^p + |\operatorname{adj} F|^q + (\det F)^r)$$

for all $\underset{\sim}{X} \in \Omega$, $F \in M^{3\times 3}$ with $\det F \geq d$.

Let ψ be C^1, and if $p < 3$ assume that there exist constants C_1 and γ, $1 \leqslant \gamma \leqslant \frac{3p}{3-p}$, $\gamma \geqslant 1$ arbitrary if $p = 3$, such that

$$\left|\frac{\partial\psi}{\partial \underset{\sim}{x}}\right| \leqslant C_1 (1 + |\underset{\sim}{x}|^\gamma)$$

for all $\underset{\sim}{x} \in \mathbb{R}^3$.

Theorem 5.1 Let $\underset{\sim}{x}_o$ be the minimizer of Theorem 4.1. Suppose that $\det \nabla \underset{\sim}{x}_o(\underset{\sim}{X}) \geqslant d_1 > 0$ almost everywhere in some open subset E of Ω. Then $\underset{\sim}{x} = \underset{\sim}{x}_o$ satisfies the Euler-Lagrange equation

$$\int_E \left[\frac{\partial\psi}{\partial x^i} v^i + \frac{\partial W}{\partial F^i_\alpha} v^i_{,\alpha}\right] d\underset{\sim}{X} = 0 \quad \text{for all } \underset{\sim}{v} \in \mathcal{D}(E). \tag{5.1}$$

Proof. Let $\underset{\sim}{v} \in \mathcal{D}(E)$. Since $p \geqslant q \geqslant r$ it follows that $\underset{\sim}{x}_o + \varepsilon \underset{\sim}{v} \in A$ for small enough $|\varepsilon|$. Also there exists a constant d such that $\det \nabla(\underset{\sim}{x}_o + \varepsilon\underset{\sim}{v})(\underset{\sim}{X}) \geqslant d > 0$ for almost all $\underset{\sim}{X} \in E$ and all small enough $|\varepsilon|$.

We must show that $\left.\frac{d}{d\varepsilon} I(\underset{\sim}{x}_o + \varepsilon\underset{\sim}{v})\right|_{\varepsilon=0}$ exists and is given by the left hand side of (5.1). But

$$\frac{I(\underset{\sim}{x}_o + \varepsilon\underset{\sim}{v}) - I(\underset{\sim}{x}_o)}{\varepsilon} = \int_E \frac{\psi(\underset{\sim}{x}_o(\underset{\sim}{X}) + \varepsilon\underset{\sim}{v}(\underset{\sim}{X})) - \psi(\underset{\sim}{x}_o(\underset{\sim}{X}))}{\varepsilon} d\underset{\sim}{X}$$

$$+ \int_E \frac{W(\underset{\sim}{X}, \nabla\underset{\sim}{x}_o(\underset{\sim}{X}) + \varepsilon \nabla\underset{\sim}{v}(\underset{\sim}{X})) - W(\underset{\sim}{X}, \nabla\underset{\sim}{x}_o(\underset{\sim}{X}))}{\varepsilon} d\underset{\sim}{X}.$$

Using the mean value theorem, hypothesis (H4) and the dominated convergence theorem, it is easily proved that the second integral tends to

$$\int_E \frac{\partial W}{\partial F^i_\alpha}(\underset{\sim}{X}, \nabla\underset{\sim}{x}_o(\underset{\sim}{X}))\, v^i_{,\alpha}(\underset{\sim}{X})\, d\underset{\sim}{X} \text{ as } \varepsilon \longrightarrow 0.$$

The first integral is treated similarly, using the facts that $\underset{\sim}{x}_o \in C(\text{supp}\,\underset{\sim}{v})$ if $p > 3$, $\underset{\sim}{x}_o \in L^{3p/(3-p)}(\text{supp}\,\underset{\sim}{v})$ if $p < 3$, $\underset{\sim}{x}_o \in L^{\gamma}(\text{supp}\,\underset{\sim}{v})$ for any $\gamma > 1$ if $p = 3$.

□

No regularity theory seems to be known for nonlinear elliptic systems of the type encountered in elasticity. Such a theory would probably require strong growth conditions on W such as (1.20) to prevent the formation of 'holes'. This is certainly what is indicated by an example due to Giusti and Miranda [15] of an analytic integrand $f(\underset{\sim}{X},\underset{\sim}{x},\nabla\underset{\sim}{x})$ with $f(\underset{\sim}{X},\underset{\sim}{x},\cdot)$ convex, such that the function

$$\underset{\sim}{x}_o(\underset{\sim}{X}) = \frac{\underset{\sim}{X}}{|\underset{\sim}{X}|},\ \underset{\sim}{X} \in \Omega,$$

Ω a bounded open subset of $\mathbb{R}^n$, $n \geq 3$, is a solution of the Euler-Lagrange equations for

$$J(\underset{\sim}{x}) = \int_{\Omega} f(\underset{\sim}{X},\underset{\sim}{x},\nabla\underset{\sim}{x})d\underset{\sim}{X}.$$

For large enough n, $\underset{\sim}{x}_o(\underset{\sim}{X})$ is in fact the unique minimizer for J subject to $\underset{\sim}{x} = \underset{\sim}{x}_o$ on $\partial\Omega$. In Giusti and Miranda's example f is quadratic in $\nabla\underset{\sim}{x}$, so that (1.20) is not satisfied.

Let us say that a stored-energy function $W(\underset{\sim}{X},F)$ is of <u>slow growth</u> if (1.20) is not satisfied. W may be of slow growth and still satisfy the hypotheses of Theorem 4.1. However, it is possible to treat other such stored-energy functions by refining the sequential weak continuity results of Section 3. We first observe that the right hand sides of (2.4), (2.5) can have meaning as distributions when the conditions of Lemma 3.3 are not satisfied. To be precise, let $\underset{\sim}{x} \in W^{1,\frac{3}{2}}(\Omega)$; then $\underset{\sim}{x} \in L^3_{loc}(\Omega)$ by the imbedding theorems, and hence $x^j x^k_{,\gamma} \in L^1_{loc}(\Omega)$. Thus (note the capital letter)

$$(\text{Adj}\,\nabla\underset{\sim}{x})^{\alpha}_{i} \overset{\text{def}}{=} (\tfrac{1}{2}\varepsilon_{ijk}\,\varepsilon^{\alpha\beta\gamma}\,x^j x^k_{,\gamma})_{,\beta}$$

is a well defined distribution. Similarly, if $\underset{\sim}{x} \in W^{1,p}(\Omega)$ and $\text{Adj}\,\nabla\underset{\sim}{x} \in L^q(\Omega)$ for $p > 1$, $q > 1$, $\frac{1}{p} + \frac{1}{q} \leqslant \frac{4}{3}$, then

$$\text{Det}\,\nabla\underset{\sim}{x} \overset{\text{def}}{=} \left(\frac{1}{3} x^i (\text{Adj}\,\nabla\underset{\sim}{x})_i^\alpha\right)_{,\alpha}$$

is a well defined distribution.

Lemma 3.3 says that if $\underset{\sim}{x} \in W^{1,2}(\Omega)$ then $\text{adj}\,\nabla\underset{\sim}{x} = \text{Adj}\,\nabla\underset{\sim}{x}$, and if $\underset{\sim}{x} \in W^{1,p}(\Omega)$, $\text{adj}\,\nabla\underset{\sim}{x} \in L^{p'}(\Omega)$, $p \geqslant 2$, then $\det\nabla\underset{\sim}{x} = \text{Det}\,\nabla\underset{\sim}{x}$. It is not always true that $\text{adj}\,\nabla\underset{\sim}{x} = \text{Adj}\,\nabla\underset{\sim}{x}$, $\det\nabla\underset{\sim}{x} = \text{Det}\,\nabla\underset{\sim}{x}$. As an example, consider the map

$$\underset{\sim}{x}(\underset{\sim}{X}) = (1 + |\underset{\sim}{X}|)\frac{\underset{\sim}{X}}{|\underset{\sim}{X}|} \quad \text{for } |\underset{\sim}{X}| < 1.$$

This map produces a spherical hole of radius 1 at $\underset{\sim}{X} = \underset{\sim}{0}$. One can check (see [A]) that $\underset{\sim}{x} \in W^{1,p}(\Omega)$ for $1 \leqslant p < 3$, $\text{adj}\,\nabla\underset{\sim}{x} \in L^q(\Omega)$ for $1 \leqslant q < \frac{3}{2}$. But $\text{Det}\,\nabla\underset{\sim}{x} \neq \det\nabla\underset{\sim}{x}$ since $\text{Det}\,\nabla\underset{\sim}{x}$ has an atom of measure $\frac{4\pi}{3}$ at $\underset{\sim}{X} = \underset{\sim}{0}$.

Open problem

6. Need $\det\nabla\underset{\sim}{x} = \text{Det}\,\nabla\underset{\sim}{x}$ if $\text{Det}\,\nabla\underset{\sim}{x}$ is a function (and a similar question for $\text{Adj}\,\nabla\underset{\sim}{x}$)?

It is obvious that the arguments of Theorem 3.4 carry over to the distributions $\text{Adj}\,\nabla\underset{\sim}{x}$, $\text{Det}\,\nabla\underset{\sim}{x}$ under weaker conditions on p. Thus one can prove the existence of minimizers for functionals of the form

$$\bar{I}(\underset{\sim}{x}) = \int_\Omega \psi(\underset{\sim}{x})d\underset{\sim}{X} + \int_\Omega G(\underset{\sim}{X}, \nabla\underset{\sim}{x}, \text{Adj}\,\nabla\underset{\sim}{x}, \text{Det}\,\nabla\underset{\sim}{x})d\underset{\sim}{X}$$

with $G(\underset{\sim}{X},\cdot,\cdot,\cdot)$ convex, under coercivity conditions on G weaker than the corresponding conditions on g in (H2). The reader is referred to [A] for details of these results. The minimization is carried out in a class of functions $\underset{\sim}{x}$ such that $\text{Adj}\,\nabla\underset{\sim}{x}$ and $\text{Det}\,\nabla\underset{\sim}{x}$ are functions, but the relationship of the integrand $G(\underset{\sim}{X}, \nabla\underset{\sim}{x}, \text{Adj}\,\nabla\underset{\sim}{x}, \text{Det}\,\nabla\underset{\sim}{x})$ to $W(\underset{\sim}{X},F) = G(\underset{\sim}{X},F, \text{adj}\,F, \det F)$ is unclear because the open problem 6 is unsolved.

We now investigate to what extent the hypotheses of our existence theorems are satisfied by accurate models of real elastic materials. We confine attention to <u>isotropic</u> materials. For an isotropic material W has the form

$$W(\underset{\sim}{X},F) = \Phi(\underset{\sim}{X},v_1,v_2,v_3), \tag{6.1}$$

where v_i are the eigenvalues of $\sqrt{FF^T}$ and where Φ is symmetric in the v_i. Because the transformation $F \longmapsto (v_1,v_2,v_3)$ is nonlinear, it is by no means obvious under what conditions on Φ the stored-energy function W is quasiconvex or polyconvex. The corollary to the following result gives some sufficient conditions. For the proof of the theorem see Thompson and Freede [36] and [A].

<u>Theorem 6.1</u> Let $\psi(v_1,v_2,v_3)$ be a symmetric real valued function defined on $\mathbb{R}_3^+ = \{v_i \geqslant 0\}$. For $F \in M^{3\times 3}$ define

$$\sigma(F) = \psi(v_1,v_2,v_3),$$

where the v_i are the eigenvalues of $\sqrt{FF^T}$. Then σ is convex on $M^{3\times 3}$ if and only if ψ is convex and nondecreasing in each variable v_i.

<u>Corollary 6.2</u> For $j = 1,2$ let $\Phi_j : \Omega \times \mathbb{R}_3^+ \longrightarrow \mathbb{R}$ be continuous and such that $\Phi_j(\underset{\sim}{X},\cdot,\cdot,\cdot)$ is symmetric, convex and nondecreasing for each $\underset{\sim}{X} \in \Omega$. Let $\Phi_3 : \Omega \times (0,\infty) \longrightarrow \mathbb{R}$ be continuous and such that $\Phi(\underset{\sim}{X},\cdot)$ is convex for each $\underset{\sim}{X} \in \Omega$.

Let

$$\Phi(\underset{\sim}{X},v_1,v_2,v_3) = \Phi_1(\underset{\sim}{X},v_1,v_2,v_3) + \Phi_2(\underset{\sim}{X},v_2v_3,v_3v_1,v_1v_2) + \Phi_3(\underset{\sim}{X},v_1v_2v_3).$$

Then W, defined by (6.1), is polyconvex.

<u>Proof.</u> By the theorem $\Phi_1(\underset{\sim}{X},v_1,v_2,v_3)$ is convex in F for each $\underset{\sim}{X}$. Also, since the eigenvalues of $\sqrt{(\text{adj } F)(\text{adj } F)^T}$ are v_2v_3, v_3v_1, v_1v_2 it follows from the theorem that $\Phi_2(\underset{\sim}{X},v_2v_3,v_3v_1,v_1v_2)$ is a convex function of adj F. Since $\det F = v_1v_2v_3$ we obtain the result. □

We consider a slight modification of a class of stored-energy functions introduced by Ogden [25]. For $\alpha \geqslant 1, \beta \geqslant 1$, let

$$\rho(\alpha) = v_1^\alpha + v_2^\alpha + v_3^\alpha - 3, \quad \chi(\beta) = (v_2v_3)^\beta + (v_3v_1)^\beta + (v_1v_2)^\beta - 3.$$

Consider first incompressible materials, and let

$$W(\underset{\sim}{X},F) = \sum_{i=1}^{M} a_i(\underset{\sim}{X})\rho(\alpha_i) + \sum_{j=1}^{M} b_j(\underset{\sim}{X})\chi(\beta_j), \tag{6.2}$$

where $\alpha_1 \geqslant \ldots \geqslant \alpha_M \geqslant 1$, $\beta_1 \geqslant \ldots \geqslant \beta_N \geqslant 1$, and where a_i, b_j are continuous functions on $\bar{\Omega}$ satisfying

$$a_i(\underset{\sim}{X}) \geqslant 0,\ b_j(\underset{\sim}{X}) \geqslant 0, \quad \text{for } 1 \leqslant i \leqslant M,\ 1 \leqslant j \leqslant N,\ \underset{\sim}{X} \in \bar{\Omega},$$
$$a_1(\underset{\sim}{X}) \geqslant k > 0,\ b_1(\underset{\sim}{X}) \geqslant k > 0, \quad \text{for } \underset{\sim}{X} \in \bar{\Omega},$$

where k is some constant.

By Corollary 6.2, W is polyconvex, the continuity of W following from the convexity of $\rho(\alpha_i)$, $\chi(\beta_j)$ as functions of F, adj F respectively. (The special case of Theorem 6.1 used here is due to von Neumann [40]). The continuity of $\rho(\alpha_1)$, $\chi(\beta_1)$ also implies the existence of positive constants a,b such that

$$\rho(\alpha_1) \geqslant a|F|^{\alpha_1}, \ \chi(\beta_1) \geqslant b|\text{adj } F|^{\beta_1}$$

for all $F \in M^{3\times 3}$. Therefore hypotheses (H1)', (H2)' of Theorem 4.2 are satisfied, provided

$$\alpha_1 \geqslant 2, \ \beta_1 \geqslant \frac{\alpha_1}{\alpha_1 - 1} \ . \tag{6.3}$$

As a special case, consider the inhomogeneous Mooney-Rivlin material, for which $M = N = 1$, $\alpha_1 = \beta_1 = 2$, so that

$$W(\underset{\sim}{X},F) = a_1(\underset{\sim}{X})(I_B - 3) + b_1(\underset{\sim}{X})(II_B - 3),$$

where I_B and II_B are the first two principal invariants of $B = FF^T$. Clearly (6.3) is satisfied, so that the Mooney-Rivlin material is included in the existence theory. An application to buckling of a rod made of Mooney-Rivlin material is described in [A]. The incompressible Neo-Hookean material

$$W(\underset{\sim}{X},F) = a_1(\underset{\sim}{X})(I_B - 3)$$

is not covered by Theorem 4.2. To illustrate this, consider the single-term stored-energy function

$$W(F) = \rho(\alpha_1).$$

It is not hard to show that $(H_2)'$ is satisfied if and only if $\alpha_1 \geqslant 3$. By use of Adj and Det one can reduce α_1 to $2\frac{1}{4}$, but this still does not cover the Neo-Hookean material.

Ogden curve-fitted a stored-energy function of the form (6.2) with three terms ($M = 2$, $N = 1$) to data of Treloar for homogeneous vulcanized rubber. The values of the various constants obtained by him were

$$\alpha_1 = 5.0, \quad \alpha_2 = 1.3, \quad \beta_1 = 2,$$

$$a_1 = 2.4 \times 10^{-3}, \quad a_2 = 4.8, \quad b_1 = 0.05 \text{ kg cm}^{-2}.$$

Similar values are given by Jones and Treloar [21]. Clearly (6.3) is satisfied. Furthermore, since $\alpha_1 > 3$, condition (1.20) holds. If $\partial\Omega$ satisfies a strong Lipschitz condition then by the imbedding theorem of Morrey [23] the minimizer in Theorem 4.2 belongs to $C^{0,0.4}(\bar{\Omega})$.

In the compressible case Ogden [26] considered the effect of adding a term $\Gamma(\det F)$ to (6.2). (Actually he replaced the term $\chi(2)$ by $v_1^{-2} + v_2^{-2} + v_3^{-2} - 3$, but the difference is negligible experimentally since for rubber $v_1 v_2 v_3 \approx 1$). Suppose that

$$\Gamma(t) \geqslant c + d\, t^r \quad \text{for all } t > 0,$$

where $d > 0$, $r > 1$ and c are constants. Assume that Γ is convex on $(0,\infty)$, $\Gamma(t) \longrightarrow \infty$ as $t \longrightarrow 0+$. Then the modified stored-energy function satisfies hypotheses (H1) - (H3) of Theorem 4.1, provided (6.3) holds.

Open problem

7. Find necessary and sufficient conditions on Φ for $W(\underset{\sim}{X},F)$ to be polyconvex.

A generalization of Corollary 6.2 is given in [A], but it does not solve the above problem.

7 EXISTENCE OF SEMI-INVERSE SOLUTIONS

In this section we prove the existence of minimizers for semi-inverse problems of the type discussed by Ericksen in [14] and in his article in this volume.†

Let D be a bounded open subset of $\mathbb{R}^2$. We denote coordinates in $\mathbb{R}^2$ by X^Γ, $\Gamma = 1,2$. Consider an elastic body which in a reference configuration occupies the cylindrical region $D \times (0,1)$. See Figure 8.

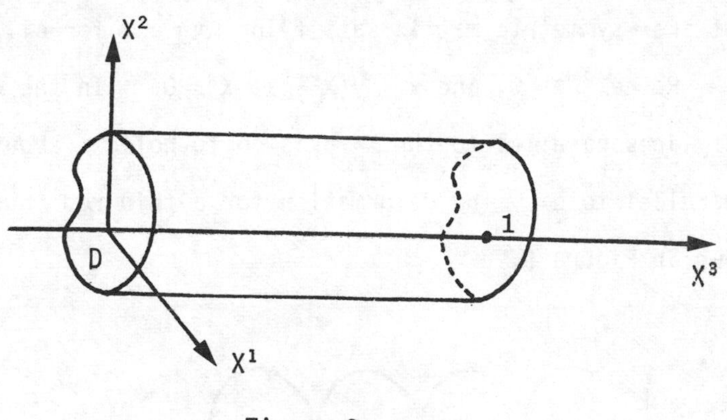

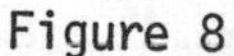

Figure 8

We make the following assumptions on W.

(A1) W is independent of X^3, so that $W : D \times U \longrightarrow \mathbb{R}$.

(A2) W is frame-indifferent (see (1.12)).

(A3) W is polyconvex and the corresponding function $g : D \times E \longrightarrow \mathbb{R}$ is continuous.

†I would like to thank Professor Ericksen for some useful discussions concerning the material in this section.

(A4) There exist constants $K > 0$, C, $p \geqslant 2$, $q > 1$, such that

$$g(X^\Gamma, F, H, \delta) \geqslant C + K(|F|^p + |H|^q)$$

for all $(X^1, X^2) \in D$, $(F, H, \delta) \in E$.

(A5) $g(X^\Gamma, a) \longrightarrow \infty$ as $a \longrightarrow \partial E$.

For simplicity suppose that there is no body force. The semi-inverse solutions have the form

$$\underset{\sim}{x}(\underset{\sim}{X}) = \underset{\sim}{k} + R[\underset{\sim}{y}(X^\Gamma) - \underset{\sim}{k} + \alpha X^3 \underset{\sim}{e}], \tag{7.1}$$

where $\underset{\sim}{e} \in \mathbb{R}^3$ is a unit vector, $\underset{\sim}{k} \in \mathbb{R}^3$, $R(X^3) = e^{\gamma\Omega X^3}$, α, γ are constants, and Ω is a constant skew-symmetric matrix satisfying $\Omega\underset{\sim}{v} = \underset{\sim}{e} \wedge \underset{\sim}{v}$ for all $\underset{\sim}{v} \in \mathbb{R}^3$. Clearly $R^1 = \gamma\Omega R$, $R\underset{\sim}{e} = \underset{\sim}{e}$, $\Omega\underset{\sim}{e} = \underset{\sim}{0}$, and $\underset{\sim}{x} = \underset{\sim}{y}(X^\Gamma)$ at $X^3 = 0$. In the deformation (7.1), straight lines parallel to the X^3 axis go to helices about the line through $\underset{\sim}{x} = \underset{\sim}{k}$ parallel to $\underset{\sim}{e}$. The deformation for a thin rod thus looks roughly as shown in Figure 9.

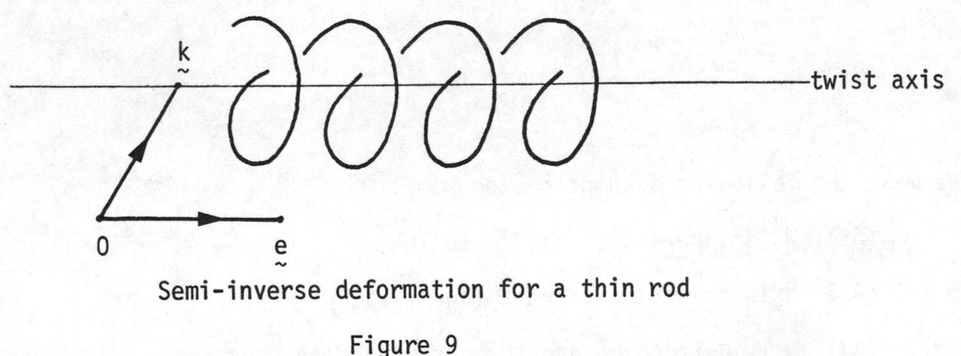

Semi-inverse deformation for a thin rod

Figure 9

If $\underset{\sim}{x}$ has the form (7.1) then $\nabla\underset{\sim}{x} = R(\nabla\underset{\sim}{y},\underset{\sim}{z})$, where $\underset{\sim}{z} = \alpha\underset{\sim}{e} + \gamma\underset{\sim}{e}\wedge(\underset{\sim}{y}-\underset{\sim}{k})$. Hence by (A2)

$$W(X^{\Gamma},\nabla\underset{\sim}{x}) = W(X^{\Gamma},\nabla\underset{\sim}{y},\underset{\sim}{z}). \tag{7.2}$$

Note that

$$\det \nabla\underset{\sim}{x} = \underset{\sim}{z} \cdot (\underset{\sim}{y}_{,1} \wedge \underset{\sim}{y}_{,2}). \tag{7.3}$$

Consider the functional

$$J(\underset{\sim}{y}) = \int_D W(X^{\Gamma},\nabla\underset{\sim}{y},\underset{\sim}{z})dS, \qquad dS = dX^1 dX^2. \tag{7.4}$$

The Euler-Lagrange equations for (7.4) are

$$\frac{\partial W}{\partial \underset{\sim}{y}} - \left(\frac{\partial W}{\partial \underset{\sim}{y}_{,\Gamma}}\right)_{,\Gamma} = \underset{\sim}{0}. \tag{7.5}$$

It is easily verified using (A1), (A2) that if (7.5) holds then, formally, so do the equilibrium equations

$$\frac{\partial}{\partial X^{\alpha}}\left(\frac{\partial W}{\partial x^{i}_{,\alpha}}\right) = 0, \tag{7.6}$$

where $\underset{\sim}{x}$ is given by (7.1). We therefore consider the problem of minimizing $J(\underset{\sim}{y})$. We will only consider the case when the curved surface of the cylinder is traction free; this is a natural boundary condition, so that for the purposes of minimization it can be ignored. Let $A = \{\underset{\sim}{y} \in W^{1,p}(D)$: $\operatorname{adj}(\nabla\underset{\sim}{y},\underset{\sim}{0}) \in L^q(D)$, $\underset{\sim}{z} \cdot (\underset{\sim}{y}_{,1} \wedge \underset{\sim}{y}_{,2}) > 0$ for almost all $(X^1,X^2) \in D\}$.

Theorem 7.1 Let D satisfy the cone condition. Let $\underset{\sim}{e},\underset{\sim}{k},\alpha$ and γ be fixed, and assume $\alpha^2 + \gamma^2 \neq 0$. Suppose there exists $\underset{\sim}{y}_1 \in A$ with $J(\underset{\sim}{y}_1) < \infty$. Then there exists $\underset{\sim}{y}_o \in A$ which minimizes J on A.

To prove the theorem we need the following version of the Poincaré inequality:

Lemma 7.2 Let D satisfy the cone condition, let $p > 1$, and let $\underset{\sim}{e} \in \mathbb{R}^3$ be a unit vector. Then the inequality

$$\int_D |\underset{\sim}{y}|^p \, dS \leqslant \text{const.}[(\int_D \underset{\sim}{y}.\underset{\sim}{e} \, dS)^p + \int_D |\underset{\sim}{e} \wedge \underset{\sim}{y}|^p \, dS + \int_D |\nabla \underset{\sim}{y}|^p \, dS] \tag{7.7}$$

holds for all $\underset{\sim}{y} : D \longrightarrow \mathbb{R}^3$ belonging to $W^{1,p}(D)$.

Proof of lemma. We use a similar argument to Morrey [23 p 82]. Since D satisfies the cone condition and is bounded, it has only finitely many connected components. Hence without loss of generality we may assume D to be connected. Suppose the lemma is false. Then there exists a sequence $\{\underset{\sim}{y}_N\} \subset W^{1,p}(D)$ such that $\int_D |\underset{\sim}{y}_N|^p \, dS = 1$ and

$$\int_D |\nabla \underset{\sim}{y}_N|^p \, dS + \int_D |\underset{\sim}{e} \wedge \underset{\sim}{y}_N|^p \, dS + (\int_D \underset{\sim}{y}_N.\underset{\sim}{e} dS)^p \leqslant \frac{1}{N}. \tag{7.8}$$

Thus $\underset{\sim}{y}_N$ is bounded in $W^{1,p}(D)$, so that a subsequence $\underset{\sim}{y}_\mu$ converges weakly in $W^{1,p}(D)$ to some $\underset{\sim}{y}$. Since the imbedding of $W^{1,p}(D)$ into $L^p(D)$ is compact, it follows that $\underset{\sim}{y}_\mu \longrightarrow \underset{\sim}{y}$ in $L^p(D)$. Hence

$$\int_D |\underset{\sim}{y}|^p \, dS = 1, \tag{7.9}$$

and by (7.8) and the fact that $\int_D |\nabla \underset{\sim}{y}|^p \, dS \leqslant \varliminf_{\mu\to\infty} \int_D |\nabla \underset{\sim}{y}_\mu|^p \, dS$, we have that

$$\nabla \underset{\sim}{y} = \underset{\sim}{0}, \ \underset{\sim}{e} \wedge \underset{\sim}{y} = \underset{\sim}{0} \text{ for almost all } (X^1, X^2) \in D,$$

and

$$\int_D \underset{\sim}{e} . \underset{\sim}{y} \, dS = 0.$$

Since D is connected, $\underset{\sim}{y}$ is constant, and hence $\underset{\sim}{y} = \underset{\sim}{0}$. This contradicts (7.9).

Proof of theorem. Suppose first that $\gamma \neq 0$. Then $W(X^\Gamma, \nabla \underset{\sim}{y}, \underset{\sim}{z})$ is invariant under the transformation $\underset{\sim}{y} \longmapsto \underset{\sim}{y} + \lambda \underset{\sim}{e}$. Hence we may without loss of generality seek a minimum for J on the set $\bar{A} = \{\underset{\sim}{y} \in A : \int_D \underset{\sim}{y} . \underset{\sim}{e} \, dS = 0\}$. Let $\underset{\sim}{y}_{(n)}$ be a minimizing sequence for J in $\bar{A}$. Then by (A4)

$$\int_D |\nabla \underset{\sim}{y}_{(n)}|^p \, dS \leq \text{constant}, \quad \int_D |\underset{\sim}{e} \wedge \underset{\sim}{y}_{(n)}|^p \, dS \leq \text{constant},$$

$$\int_D |\operatorname{adj}(\nabla \underset{\sim}{y}_{(n)}, \underset{\sim}{z}_{(n)})|^q \, dS \leq \text{constant}, \quad \text{where } \underset{\sim}{z}_{(n)} = \alpha \underset{\sim}{e} + \gamma \underset{\sim}{e} \wedge (\underset{\sim}{y}_{(n)} - \underset{\sim}{k}).$$

Hence by the lemma $\|\underset{\sim}{y}_{(n)}\|_{W^{1,p}(D)} \leq$ constant, and so there exists a subsequence $\underset{\sim}{y}_{(\mu)}$ of $\underset{\sim}{y}_{(n)}$ satisfying

$$\underset{\sim}{y}_{(\mu)} \longrightarrow \underset{\sim}{y}_o \text{ in } W^{1,p}(D), \quad \underset{\sim}{y}_\mu \longrightarrow \underset{\sim}{y}_o \text{ in } L^r(D) \text{ for any } r > 1,$$

$$\operatorname{adj}(\nabla \underset{\sim}{y}_{(\mu)}, \ \underset{\sim}{z}_{(\mu)}) \longrightarrow H \text{ in } L^q(D).$$

Arguments similar to those in Section 3 show that $H = \operatorname{adj}(\nabla \underset{\sim}{y}_o, \underset{\sim}{z}_o)$, where $\underset{\sim}{z}_o = \alpha \underset{\sim}{e} + \gamma \underset{\sim}{e} \wedge (\underset{\sim}{y}_o - \underset{\sim}{k})$. Hence also

$$\underset{\sim}{z}_{(\mu)} \cdot (\underset{\sim}{y}_{(\mu),1} \wedge \underset{\sim}{y}_{(\mu),2}) = z^i_{(\mu)} \operatorname{adj}(\nabla \underset{\sim}{y}_{(\mu)}, \underset{\sim}{0})^3_i$$

converges weakly to $\underset{\sim}{z}\cdot(\underset{\sim}{y}_{0,1}\wedge\underset{\sim}{y}_{0,2})$ in $L^1(D)$. Hence by the same argument as in Theorem 4.1 we find that $\underset{\sim}{y}_o \in \bar{A}$ and $J(\underset{\sim}{y}_o) = \inf_{\bar{y}\in\bar{A}} J(\bar{\underset{\sim}{y}})$.

If $\gamma = 0$, $\alpha \neq 0$, then $W(X^\Gamma, \nabla\underset{\sim}{y}, \underset{\sim}{z})$ is invariant under the transformation $\underset{\sim}{y} \longmapsto \underset{\sim}{y} + \underset{\sim}{a}$. Hence it is sufficient to minimize J on the set $\bar{\bar{A}} = \{\underset{\sim}{y} \in A : \int_D \underset{\sim}{y}\, dS = \underset{\sim}{0}\}$. This is done in the same way as for $\gamma \neq 0$ by using the Poincaré inequality

$$\int_D |\underset{\sim}{y}|^p\, dS \leqslant \text{const.}[|\int_D \underset{\sim}{y}\, dS|^p + \int_D |\nabla\underset{\sim}{y}|^p\, dS]$$

for all $\underset{\sim}{y} \in W^{1,p}(D)$. □

Note that because D is two-dimensional we get existence under the growth condition (A4), which is weaker than the corresponding hypothesis (H2) for the full three-dimensional problem.

References

[A] J. M. Ball, Convexity conditions and existence theorems in non-linear elasticity, Arch. Rat. Mech. Anal. 63 (1977) 337-403.

1 R. A. Adams, "Sobolev spaces", Academic Press, New York, 1975.

2 S. S. Antman, Existence of solutions of the equilibrium equations for nonlinearly elastic rings and arches, Indiana Univ. Math. J. 20 (1970) 281-302.

3 S. S. Antman, Monotonicity and invertibility conditions in one-dimensional nonlinear elasticity, in "Nonlinear Elasticity", ed. R. W. Dickey, Academic Press, New York, 1973.

4 S. S. Antman, Ordinary differential equations of nonlinear elasticity II: Existence and regularity theory for conservative boundary value problems, Arch. Rat. Mech. Anal. 61 (1976) 353-393.

5 J. M. Ball, On the calculus of variations and sequentially weakly continuous maps, in "Ordinary and Partial Differential Equations", Dundee 1976, Springer Lecture Notes in Mathematics, Vol. 564, 13-25.

6 H. Busemann, G. Ewald and G. S. Shephard, Convex bodies and convexity on Grassman cones, Parts I - IV, Math. Ann. 151 (1963) 1-41.

7 H. Busemann and G. S. Shephard, Convexity on nonconvex sets, Proc. Coll. on Convexity, Copenhagen, Univ. Math. Inst., Copenhagen, (1965) 20-33.

8 B. D. Coleman and W. Noll, On the thermostatics of continuous media, Arch. Rat. Mech. Anal., 4 (1959) 97-128.

9 D. G. B. Edelen, The null set of the Euler-Lagrange operator, Arch. Rat. Mech. Anal., 11 (1962) 117-121.

10 I. Ekeland and R. Témam, "Analyse convexe et problèmes variationnels", Dunod, Gauthier-Villars, Paris, 1974.

11 J. L. Ericksen, Nilpotent energies in liquid crystal theory, Arch. Rat. Mech. Anal., 10 (1962) 189-196.

12 J. L. Ericksen, Loading devices and stability of equilibrium, in "Nonlinear Elasticity", ed. R. W. Dickey, Academic Press, New York 1973.

13 J. L. Ericksen, Equilibrium of bars, J. of Elasticity, 5 (1975) 191-201.

14 J. L. Ericksen, Special topics in elastostatics, Advances in Applied Mechanics (to appear).

15 E. Giusti and M. Miranda, Un esempio disoluzioni discontinue per un problema di minimo relativo ad un integrale regolare del calcola delle variazioni, Boll. Unione Mat. Ital. Ser. 4, 1 (1968) 219-226.

16 L. M. Graves, The Weierstrass condition for multiple integral variation problems, Duke Math. J., 5 (1939) 656-660.

17 J. Hadamard, Sur une question de calcul des variations, Bull. Soc. Math. France, 30 (1902) 253-256.

18 J. Hadamard, "Lecons sur la propagation des ondes", Paris, Hermann, 1903.

19 F. John, Remarks on the non-linear theory of elasticity, Semin. Ist. Naz. Alta Mat., 1962-3, 474-482.

20 F. John, Uniqueness of non-linear elastic equilibrium for prescribed boundary displacements and sufficiently small strains. Comm. Pure Appl. Maths., 25 (1972) 617-634.

21 D. F. Jones and L. R. G. Treloar, The properties of rubber in pure homogeneous strain, J. Phys. D. Appl. Phys., 8 (1975) 1285-1304.

22 C. B. Morrey, Quasiconvexity and the lower semicontinuity of multiple integrals, Pacific J. Math., 2 (1952) 25-53.

23 C. B. Morrey, "Multiple integrals in the calculus of variations", Springer, Berlin, 1966.

24 R. W. Ogden, Compressible isotropic elastic solids under finite strain-constitutive inequalities, Quart. J. Mech. Appl. Math. 23 (1970) 457-468.

25 R. W. Ogden, Large deformation isotropic elasticity - on the correlation of theory and experiment for incompressible rubberlike solids, Proc. Roy. Soc. London A 32 (1972) 565-584.

26 R. W. Ogden, Large deformation isotropic elasticity: on the correlation of theory and experiment for compressible rubberlike solids, Proc. Roy. Soc. London A (1972) 567-583.

27 Y. G. Reshetnyak, On the stability of conformal mappings in multidimensional spaces, Sibirskii Math. 8 (1967) 91-114.

28 Y. G. Reshetnyak, Stability theorems for mappings with bounded excursion, Sibirskii Math., 9 (1968) 667-684.

29 R. S. Rivlin, Large elastic deformations of isotropic materials II. Some uniqueness theorems for pure homogeneous deformation, Phil. Trans. Roy. Soc. London 240 (1948) 491-508.

30 R. S. Rivlin, Some restrictions on constitutive equations, Proc. Int. Symp. on the Foundations of Continuum Thermodynamics, Bussaco 1973.

31 R. S. Rivlin, Stability of pure homogeneous deformations of an elastic cube under dead loading, Quart. Appl. Math., 32 (1974) 265-272.

32 R. T. Rockafellar, "Convex analysis", Princeton University Press, Princeton, New Jersey, 1970.

33 L. Schwartz, "Théorie des distributions", Hermann, Paris, 1966.

34 E. Silverman, Strong quasi-convexity, Pacific J. Math., 46 (1973) 549-554.

35 F. Stoppelli, Un teorema di esistenza e di unicita relativo alla equazioni dell'elastostatica isoterma per deformazioni finite, Ricerche Matematica, 3 (1954) 247-267.

36 R. C. Thompson and L. J. Freede, Eigenvalues of sums of Hermitian matrices III, J. Research Nat. Bureau of Standards B, 75B (1971) 115-120.

37 C. Truesdell, The main open problem in the finite theory of elasticity (1955), reprinted in "Foundations of Elasticity Theory", Intl. Sci. Rev. Ser. New York: Gordon and Breach 1965.

38 C. Truesdell and W. Noll, "The non-linear field theories of mechanics", in Handbuch der Physik Vol. III/3, ed. S. Flügge, Springer, Berlin, 1965.

39 W. van Buren, "On the existence and uniqueness of solutions to boundary value problems in finite elasticity", Thesis, Department of Mathematics, Carnegie-Mellon University, 1968. Research report 68-ID7-MEKMA-RI, Westinghouse Research Laboratories, Pittsburgh, Pa., 1968.

40 J. von Neumann, Some matrix-inequalities and metrization of matric-space, Tomsk Univ. Rev. 1 (1937) 286-300. Reprinted in Collected Works Vol. IV, Pergamon, Oxford, 1962.

41 C.-C. Wang and C. Truesdell, "Introduction to rational elasticity", Noordhoff, Groningen, 1973.

Dr. J. M. Ball, Department of Mathematics, Heriot-Watt University,
Riccarton, Currie, EDINBURGH EH14 4AS, SCOTLAND.